microbial seascapes

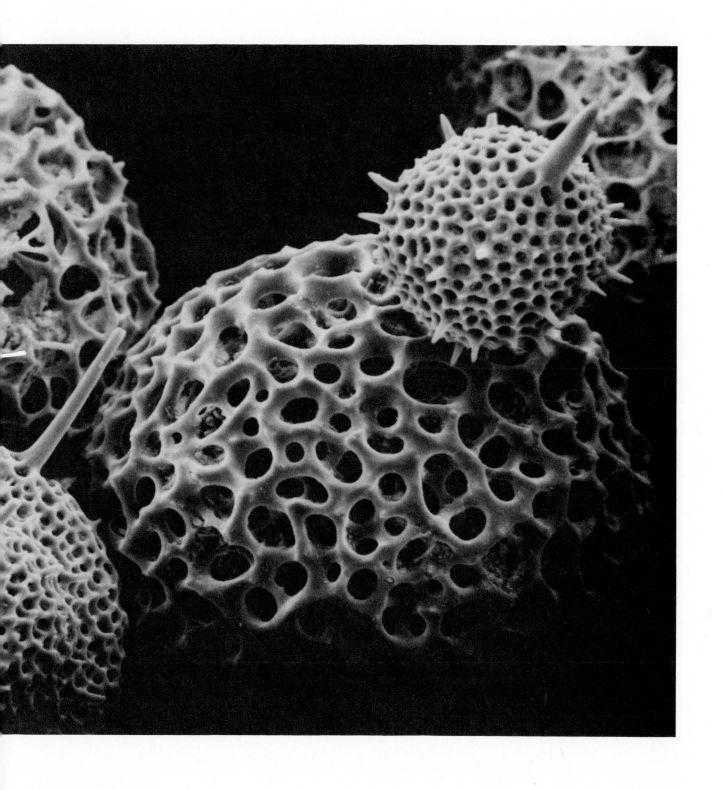

MICROBIAL SEASCAPES
A Pictorial Essay on Marine Microorganisms and Their Environments

by

John McNeill Sieburth

Professor of Oceanography and Microbiology
University of Rhode Island

with Underwater Photomacrography by
Harold L. Pratt, Biologist
National Marine Fisheries Service

and with the assistance of
Paul W. Johnson, Microbiologist
University of Rhode Island

and

Donald Scales, Specialist
University of Rhode Island

University Park Press
Baltimore ● London ● Tokyo

University Park Press
International Publishers in Science and Medicine
Chamber of Commerce Building
Baltimore, Maryland 21202

Typeset by The Composing Room of Michigan
Printed in the United States of America by Universal Lithographers, Inc.

Library of Congress Cataloging in Publication Data

Sieburth, John McNeill.
    Microbial seascapes.

    Includes bibliographical references and indexes.
    1. Marine microbiology—Pictorial works.
    I. Pratt, Harold L. II. Title.
    QR106.S53   576'.19'2   75-6935
    ISBN 0-8391-0802-8

*The Five Kingdom System diagrams appear courtesy of
the American Association for the Advancement of Science
(copyright 1969 by the American Association for the Advancement of Science),
and the comparison of microscopes figure appears courtesy of
Hamilton Publishing Company, Santa Barbara, California.*

contents

preface
acknowledgments
dedication

references
microorganism index
surface index

preface

Every submerged surface in the sea is a potential habitat for a microbial garden of epibiotic (surface-colonizing) microorganisms. These communities of microorganisms provide a source of nutrition for the larger organisms, and give a continuity to life in the sea among the sporadic blooms of microorganisms in the plankton (passively floating organisms). Until recently, the study of marine microorganisms was limited to the free-floating planktonic populations occurring in samples taken blindly, and to attached microorganisms on thin and transparent materials which could be examined by light and transmission electron microscopy. The development of satisfactory apparatus and procedures for scanning electron microscopy in the late 1960s now permits the observation of thick, opaque, and fragile surfaces and the marine microorganisms living thereon.

Skin diving, using snorkel and SCUBA, has provided the amateur and the professional naturalist with an opportunity to observe life in the sea directly. During the last three decades, diving scientists have made observations on the natural history of submerged plants and animals equivalent to those made by terrestrial naturalists during the preceding two centuries. The natural history of marine microorganisms, which require magnification for detection by the human eye, has not kept pace.

The techniques of skin diving, photomacrography, and scanning electron microscopy can be combined to observe and document natural marine surfaces, collect valid specimens, and examine them for microbial colonization. My colleagues and I have used these techniques to begin an exploration of the epibiotic microbial world growing on submerged surfaces in the sea. This volume is a collection of scanning electron micrographs of selected marine surfaces, introduced with pertinent underwater macrographs. They were obtained primarily from the shallow, temperate to polar waters of Narragansett Bay (Rhode Island Marine Bibliography, 1972) and from the subtropical waters adjacent to the Pigeon Key Marine Biological Station of the University of Miami, in the lower Florida Keys (Zischke, 1973). Oceanic materials were obtained by students and colleagues during cruises of R/V *Trident,* while other materials are from aquacultural studies conducted at the University of Rhode Island. It is hoped that this visualization of the marine microbial world will entice the underwater naturalist, the marine biologist, and the microbiologist to continue the exploration of this neglected area of study.

John McNeill Sieburth
Narragansett Marine Laboratory
Narragansett, Rhode Island

acknowledgments

I would like to sincerely acknowledge the taxpayers who sponsor my work and provide me with facilities through the University of Rhode Island and the Biological Oceanography Program of the National Science Foundation (Grant GA-41501X). My colleagues at the Bay Campus are also thanked for providing an atmosphere conducive to studying the fascinating and challenging problems discussed in this essay. Among the many individuals who have aided this project are James P. Kennett, Nelson Marshall, E. James Archer, and Dirk Frankenberg, who were instrumental in obtaining the scanning electron microscope; Judith Murphy, Susmo Honjo, and Gwendolyn Wendelschafer, who gave counsel on specimen preparation; Don Spiering, who helped with the field work at Pigeon Key; Paul Hargraves, Marie Abbott, and Donald J. Zinn, who consulted on species identification; and Jean Knapp and David Pratt, who provided valuable suggestions during the preparation of this book. Among the students whose micrographs are included in this study are W. Lloyd Balderston, Richard D. Brooks, Dennis M. Lavoie, Cynthia D. Thomas, and J. Lawton Tootle.

Permission to use the diagrams of the Five Kingdom System (Whittaker, 1969), and the comparison of microscopes (Neushul, 1974), as well as previously published micrographs (Brooks, Goos, and Sieburth, 1972; Sieburth and Thomas, 1973), is gratefully acknowledged, as is Plate 2-11 by Paul Hargraves, and Plate 2-12 by Bruce Corliss. Thierry Jacques, Gerald Levine, and James P. Kennett very kindly provided the star corals, diseased salmon, and deep sea sedimentary materials, respectively. Sheril D. Burton of Brigham Young University is especially acknowledged for his suggestion that I use my growing collection of micrographs to prepare a pictorial essay.

Snug Cove, North Vancouver, British Columbia, Canada   1932   photo by H. C. Sieburth

To my mother, Mary McNeill Sieburth,
who introduced me to the study of natural history;
and to my late father, Herman C. Sieburth,
who shared his love for the sea and photography.

part one

ENVIRONMENTS, MICROBES, AND METHODS

Plate 1-1. Surf near Flat Rock, Narragansett, RI

Microbial life in the sea starts at the sea's edge and is ubiquitous throughout its vast depths and myriad surfaces.

environments, microbes, and methods

Microbial Environments

The struggle for life in the sea, which is apparent to the fisherman, also occurs with equal vigor in the subvisible world of microorganisms. This struggle for survival has led each type of microorganism to specialize, at least to some degree, in its food preferences and the kind of habitat it seeks.

There are five major microbial environments in the sea. The first is the planktonic environment consisting of unattached microorganisms floating in the water mass. These microorganisms have received the most attention through the years, as they form the base of the food chain for the higher organisms (Parsons and Takahashi, 1973; Raymont, 1963). These microorganisms are briefly examined, along with other suspended particles such as mineral fragments, in Part Two. A second major microbial environment is the sea floor, composed of mineral fragments, skeletons of organisms, and organic debris settling out of the water column. These materials serve as a substratum or source of nutrition for microorganisms which in turn are food for animals inhabiting the sediment (Fenchel, 1969, 1970). This environment and its microorganisms are described in Part Eight.

Between these two extremes is the third major microbial environment, the epibiotic environment, consisting of the sessile and crawling microorganisms that form films on virtually every surface submerged in the sea. Plant, animal, and man-made surfaces and their microorganisms are covered in Parts Three, Four, Five, Six, and Seven. Diatoms, followed by bacteria and protozoans, seem to play a dominant role as the primary colonizers of surfaces in many microbial films (Skerman, 1956). The presence of these microbial films apparently encourages the settlement of invertebrate larvae (Crisp and Ryland, 1960; Meadows and Williams, 1963). The invertebrate populations attract juvenile and small fish species which in turn attract the larger carnivorous fish. It is the development of the fouling communities living on microbial films which helps make marinas, piers, and artificial reefs very productive fishing areas. Only recently has the fouling marine surface been recognized as a major site of energy production, especially in shallow waters. The fouling process formerly attracted attention as a continual nuisance and source of economic loss to ship owners. Studies initiated by navies and paint manufacturers traditionally have followed the fouling sequence by examining immersed glass slides with light microscopy. The very narrow depth of field of the light microscope has limited the examination of natural surfaces to thin specimens such as the fine filamentous algae (Brock, 1966) and the appendages of small crustaceans (Johnson et al., 1971). Recent developments in scanning electron microscopy allow the preparation of natural surfaces and biological specimens and the examination of their microorganisms. This, primarily, is what this book is all about.

The last two microbial environments, although very important ecologically, are only touched upon here, as they are hidden within other organisms. The fourth environment is the enteric habitat within the alimentary canal of animals. A variety of microorganisms live as commensals (closely associated but non-parasitic) on the food of animals and extend from the oral cavity and gills, through the digesting food, and are then packaged in the fecal pellets which they enrich for the coprophagous (feces-eating) species. The fifth microbial environment is within the cells of other microorganisms and within the tissues of the multicellular plants and animals. A variety of microorganisms live within marine organisms as parasites at the expense of the host and as symbionts contributing to the nutrition of the host (Sindermann, 1970; Taylor, 1973).

Helpful sources for identifying the plant and animal surfaces examined in this book are the general guides to marine plants and animals (Arnold, 1968; Zischke, 1973) or the more specific guides to seaweeds (Taylor and Villalard, 1972) and marine invertebrates (Miner, 1950). Robbins and Yentsch (1973) provide a useful guide to the New England shore.

Microbes and Microscopy

Traditionally, organisms were considered to be either plants or animals. This now defunct concept could never show the evolutionary and nutritional relationships of the higher forms to the small but biologically active and ecologically important microorganisms. A Five Kingdom concept proposed by Whittaker (1969) takes both evolutionary and nutritional relationships into consideration, as shown in Figure 1. This system is based on three levels of cellular organization with nutritional divergence at each level. The procaryotes (small cells without a membrane-bound nucleus and organelles) are included in the Kingdom Monera and lack the ingestive mode of nutrition. All other cells are eucaryotic and have a true membrane-bound nucleus and a variety of structurally complex organelles. Eucaryotic microorganisms that are unicellular are put in the Kingdom Protista. Here, the nutritional modes are continuous, with the photosynthetic mode of nutrition of the microalgae on the left, while the protozoa, with an ingestive mode of nutrition, are on the right. In the multicellular and multinucleate eucaryotic organisms, the nutritional modes lead to a more diversified cellular organization, and these eucaryotes are found in the higher three kingdoms, Plantae, Fungi, and Animalia. The microorganisms occur in three of the five kingdoms: the bacteria and the blue-green algae (Cyanophyta) in the Kingdom Monera (Procaryotae), the microalgae and protozoa in the Kingdom Protista, and the yeasts and molds in the Kingdom Fungi.

Whittaker's Five Kingdom concept is rapidly gaining recognition in contemporary books on the life sciences. One reason for its acceptance is that it takes evolutionary relationships into consideration and is compatible with recent biochemical, genetic, and ultrastructural studies which suggest

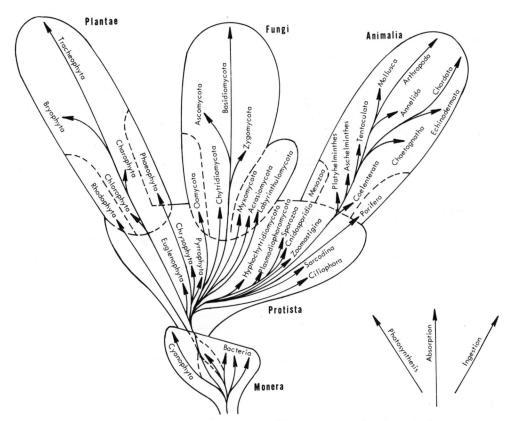

Figure 1. Whittaker's Five Kingdom concept. Reprinted by permission (Whittaker, 1969).

that hereditary endosymbiosis (living together, one inside the other) led to the eucaryotic cell of today from a variety of procaryotes that developed from a common procaryotic ancestor (Margulis, 1968, 1970). A second reason for this scheme's acceptance is that in addition to the synthetic mode of nutrition of green plants and algae and the ingestive mode of animals and protozoa, it recognizes the third mode of nutrition, absorption. Absorption cuts across the three levels of cellular organization, being the sole means of nutrition for the heterotrophic (using organic carbon as an energy source) bacteria, certain non-synthesizing and non-ingestive protists such as the apochlorotic (lacking a chlorophyll-containing photosynthetic apparatus) diatoms and flagellates, and for the multinucleated and muticellular fungi. The association of a principal nutritional mode and taxonomic affinities with the distinctive forms is important to microbial ecologists who look at microbial forms.

Microscopes used to detect and study microorganisms fall into three basic types. Each reveals a different kind of information and each has its own particular application. The similarities and differences between the three types of microscopy are shown in the simplified diagrams in Figure 2.

The light microscope and the transmission electron microscope work on the same principle, in that a thin specimen inserted within the lens system bends the path of light rays or electrons which are focused to form an image. The optical system of the light microscope uses glass lenses to refract the light passing through the whole microorganism. The light microscope is very useful for examining the larger microorganisms and their larger external and internal structures as well as the general forms of the smaller micro-organisms. The transmission electron microscope (TEM) uses electrons and magnetic lenses rather than light rays and glass lenses. The image seen with

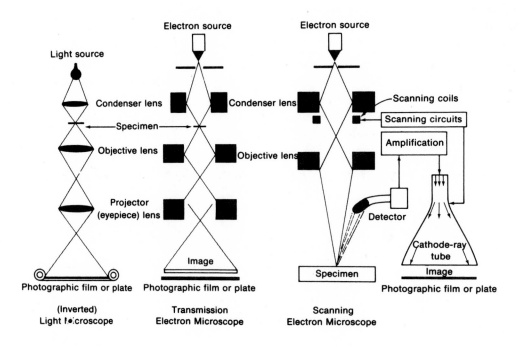

Figure 2. Comparison of three types of microscopy.

the naked eye in the light microscope or on a fluorescent screen in the TEM can be recorded on a photographic film or plate. Although whole micro-organisms can be seen in negatively stained preparations in the TEM, its main application is to use thin, stained slices (700 Å) in cross section to examine the characteristics of membranes and internal cellular organelles and even ingested particles which are observed to magnifications of 60,000X or more. These details cannot be seen in the light microscope, whose useful magnification ends at about 1,000X.

The scanning electron microscope (SEM) differs from the light microscope and the TEM. A thick, opaque specimen up to 12 mm in diameter is chemically fixed, dried, and coated with a thin film of metal. A beam of electrons focused by magnetic lenses similar to those in the TEM scans back and forth across the surface of the specimen below the lens. Electrons which are backscattered from the specimen surface are detected electronically, amplified, and displayed on a cathode ray tube like those used in television receivers. Unlike the light and transmission electron microscopes which can detect telltale internal structures, the SEM only sees the outer three-dimensional shape while the internal structures are hidden by the outer surface. This disadvantage is outweighed by the ability to observe thick, opaque surfaces and to have a four- to five-hundredfold increase in the depth of field over that of the light and transmission electron microscopes. This property makes the SEM ideal for observing the microbial seascapes formed by natural populations of marine microorganisms on marine surfaces.

Representative types of organisms from the Five Kingdoms of Whittaker are compared using the SEM and TEM in Plates 1-2 and 1-3. The TEM micrographs in Plate 1-3 show the intracellular nature of the microorganisms and their hosts, which we will be seeing only from the outside in the SEM micrographs in this book. In the SEM the multicellular and eucaryotic nature is masked by the external layer, while these properties are readily apparent in the TEM. Similarly, the large multicellular nature of a worm is unmistakable in SEM micrographs, but only TEM micrographs of a similar type of animal tissue reveal the distinctive membrane-bound nucleus and other organelles. The mycelia of fungi which could be mistaken for bacterial or algal filaments in the SEM have their own distinctive eucaryotic fine structure revealed in the TEM, and are seen to be distinct from the procaryotic bacterial filaments and the chloroplast-containing eucaryotic algae. The forms of the protists are easily identifiable in SEM micrographs but are seen to be unmistakably unicellular and eucaryotic in TEM micrographs. The distinctive procaryotic nature of the bacteria and blue-green algae seen in the TEM is only suggested by the smaller and distinctive outer form seen in the SEM.

When looking at micrographs and macrographs it is very important to know the magnification, in order to orient yourself to the object and its surroundings. The objects shown in our underwater macrographs are enlarged about 10 to 20 times. At this magnification only a few exceptionally large individual microorganisms are detectable. However, accumulations of microorganisms, such as patches of diatoms, can sometimes be seen. Although some protists such as the larger diatoms and protozoa are detectable at 50X, magnifications of 200 to 1,000X are required to see many of their details. The smaller microorganisms such as the bacteria are barely detectable at 500X and require from 1,000 to 5,000X to be seen properly. Magnifications beyond this may provide no more information and give a poorer image. Most of our micrographs are arranged in the series in which the magnification increases. The ability to detect specific types of microorganisms at certain magnifications will become quite apparent as one leafs through the different sequences of micrographs.

Methods

The use of skindiving for characterizing the physical and biological nature of shallow waters is the subject of a text edited by J. D. Woods and J. N. Lythgoe (1971). Useful hardware and procedures for surveying shallow sea bottoms by SCUBA were described by Fager et al. (1966). During the past several years Harold (Wes) Pratt has routinely spent several noonhours per week recording by macrograph the seasonal development of marine life while snorkeling in a 1-acre underwater zoological-botanical garden between the EPA Pier and the Narragansett Bay Campus small boat basin. This experience, plus the many surveys and night dives in both subtropical and temperate waters made in connection with his work at the National Marine Fisheries Service, have given Wes the experience and inspiration to select, modify, and develop useful gear for these pursuits. Plate 1-4 shows some of the gear used for the examination, photomacrography, and collection of specimens. A key item is the float, which carries the diver's safety flag and provides a buoyant platform to hang onto while resting. In addition to a safety line, which serves as an umbilical cord for a diving buddy in murky waters, the float also has watertight sample containers. The hand lens developed and described by Pratt (1973) magnifies objects to the same magnitude as the finished 8X10 macrographs. When not in use, the camera outfit and other apparatus are also suspended from the float.

All underwater photographs were taken with a 35mm format Nikonos camera using either a 35mm or 28mm primary lens. Light was provided by a Honeywell Strobonar 710 electronic flash in an Anchor Shack housing, usually with a bounce reflector opposite the strobe. The turbid waters of Narragansett Bay necessitate the attachment of closeup accessories, usually a 1:1 extension tube, which provides a finished print (8X10) with a magnification of 10 to 20X. Kodak Panatomic-X and Plus-X films were used. Typical camera settings were 1/60 sec at f/22, with distance set at minimum focus.

Sampling of representative biological specimens and marine surfaces was made after they were photographed in situ. Specimens were either drawn into inverted air-filled sterile tubes as they were righted and the air escaped, or were scooped up with their surrounding water into freshly opened Whirl-Pak containers. The samples were kept in a polyethylene jar cemented to the float for this purpose, and quickly brought to the laboratory and prepared for microscopy. Specimens were fixed in glutaraldehyde followed by osmium tetroxide to preserve the structural detail of the epibiotic microflora that would otherwise change during dehydration. All specimens must be dehydrated in order to be observed in the vacuum of the SEM microscope column. Dehydration was accomplished after fixation with a graded series of ethyl alcohol in distilled water to absolute alcohol and freeze-dried or then brought to 100% amyl acetate and critical point dried. A few of the oldest specimens were freeze-dried with an Edwards-Pearse tissue dryer Model EPD-2; however, most samples were critical point dried in liquid CO_2 in a Sorvall Critical Point Drying System. The dry specimens were mounted on SEM stubs (12.5-mm diameter aluminum discs used to hold the specimens) with plastic cement or double coated tape. The completed stubs were then coated with a thin film of gold/palladium in a Denton DV-502 high vacuum evaporator. The coated stubs were examined in a Cambridge S-4 scanning electron microscope at 10 to 20 kv. The microscope is shown in operation in Plate 1-5. Specific details and general discussions on the application of scanning electron microscopy to biological materials are covered by Echlin (1972) and Hays (1973).

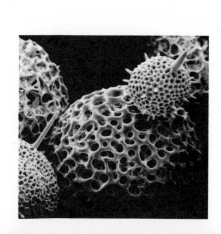

Plate 1-2. Scanning electron micrographs of the Five Kingdoms

Scanning electron micrographs of examples of organisms in the Five Kingdoms of Whittaker illustrate their differences in form and size.

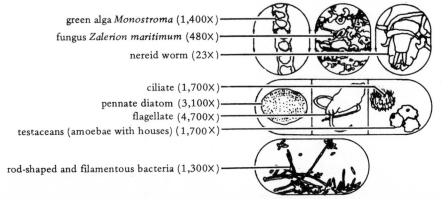

green alga *Monostroma* (1,400×)
fungus *Zalerion maritimum* (480×)
nereid worm (23×)

ciliate (1,700×)
pennate diatom (3,100×)
flagellate (4,700×)
testaceans (amoebae with houses) (1,700×)

rod-shaped and filamentous bacteria (1,300×)

Plate 1-3. Transmission electron micrographs of the Five Kingdoms

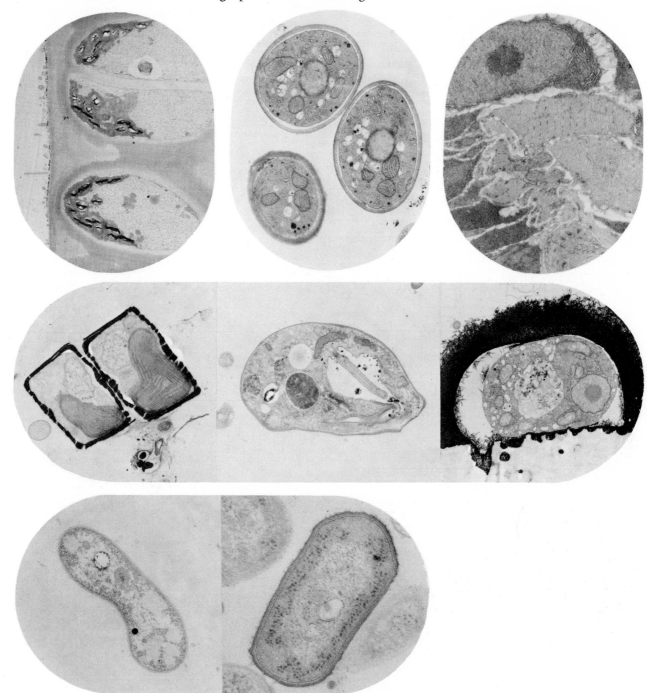

Transmission electron micrographs of thin sections of examples of organisms in the Five Kingdoms of Whittaker illustrate the differences in ultrastructure and cell organization.

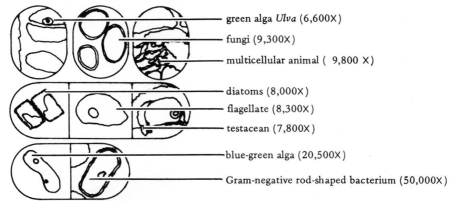

green alga *Ulva* (6,600X)

fungi (9,300X)

multicellular animal (9,800 X)

diatoms (8,000X)

flagellate (8,300X)

testacean (7,800X)

blue-green alga (20,500X)

Gram-negative rod-shaped bacterium (50,000X)

Plate 1-4. Macrophotography and collection of specimens

Wes Pratt is holding the Nikonos camera and strobe used for documenting the samples. Author is putting sample into the jar in the safety float. The underwater hand lens (Macrosnooper) by his left hand is used for examining objects at 10X magnification.

Plate 1-5. Examination of specimens with the Cambridge S-4 SEM

Johnson (left) is operating the stage controls to examine an area of interest while Scales (right) prepares to take a Polaroid micrograph. Note microscope column behind Johnson, raster line on visual display cathode ray tube on left, and camera that is in front of recording cathode ray tube on right.

part two

SUSPENDED PARTICLES

The ever-present cloud of suspended particles in rich inshore waters provides a background for a colony of the hydroid Tubularia crocea.

suspended particles

The shallow inshore waters of estuaries, continental shelves, and ocean islands are often turbid with suspended particles. This is especially true in rich temperate waters. The standing stocks of phytoplankton (floating microscopic algae) periodically and seasonally burst into blooms, increasing the turbidity and occasionally even discoloring the water. Even during these bloom conditions, the bulk of the suspended particles are not whole living cells but are fragments of these microorganisms and of other organisms and the smaller and lighter fractions of minerals held in suspension by the movement of the sea water. This organic debris and sediment, termed detritus, is an important part of the ecosystem of inshore microbial populations. Some aspects of the role of detritus in aquatic ecosystems have been considered in a symposium (Melchiorri-Santolini and Hopton, 1972).

Epibiotic microorganisms are usually quite sparse on particles suspended in the water column. Healthy populations of centric (radially symmetrical) diatoms, which often dominate the phytoplankton in temperate areas, are usually free of adhering bacteria (Droop and Elson, 1966). This is quite apparent in our micrographs. Diatoms live within beautifully structured "glass houses" (Hendey, 1959) and depend upon the transport of nutrients and waste products through the numerous pores and openings in their silicon dioxide frustules (shells). Encrustation of these pores with bacteria would interfere with their life processes. The mechanism which repels bacterial attachment is unknown, but may be as simple as the release of acidic substances from the cell (Sieburth, 1968). The organic debris suspended in the water column has also been reported to have very few attached bacteria (Jannasch, 1973; Wiebe and Pomeroy, 1972). This is also indicated in our SEM micrographs; however, bacteria may be hidden within the bacterial slime on the debris. Only serial cross sections of such material may show the true extent of bacterial development. Much of the organic debris held in suspension seems to be fecal pellets in various stages of disruption excreted by vertebrate and invertebrate animals. The feces of some copepods (planktonic crustaceans) have a covering which seems to keep them intact until they undergo decomposition on the ocean floor (Schrader, 1971). The variety and specificity of feces have been discussed by Moore (1931). Feces and their remnants are important as food for bacteria and for the coprophagous animals that live on these bacterially enriched materials (Frankenberg and Smith, 1967; Johannes and Satomi, 1966).

In the euphotic (sunlit) upper layer of the sea, the phytoplankton provide a food base for all other forms of life. The phytoplankton consist not only of diatoms, but a variety of phytoflagellates, and, in the tropics, sporadic but often large blooms of blue-green algae. The fine structure which distinguishes the many forms of flagellates has been summarized by Dodge (1973). Many of the flagellates are small and pass through the usual phytoplankton nets or, being fragile, do not survive the preparative procedures for microscopy. Some of the more resistant flagellates, such as the "armored" dinoflagellates with their outer covering of thick plates, can be seen, but usually without their flagella. A very few species of dinoflagellates under exceptional conditions are concentrated to population densities sufficient to cause "red tides" (Ryther, 1955). Some of these toxic species are concentrated in filter-feeding molluscs, including mussels and clams, causing shellfish poisoning in man (Prakash, 1962).

The phytoplankters release a portion of their synthesized organic matter in a dissolved form that supports bacterial populations, which in turn provide growth factors required by the phytoplankton. The phytoplankton and their associated bacteria are ingested by pelagic protozoa such as the foraminifera and tintinnids, which live in distinctive tests or loricae (houses). Both the phytoplankton and protozoa provide food for larval forms of larger animals and for copepods in the plankton. These herbivorous and carnivorous copepods form the basic food for many oceanic animals including fish, penguins, and the large baleen whales. Biological oceanic processes in the water column have been summarized by Parsons and Takahashi (1973).

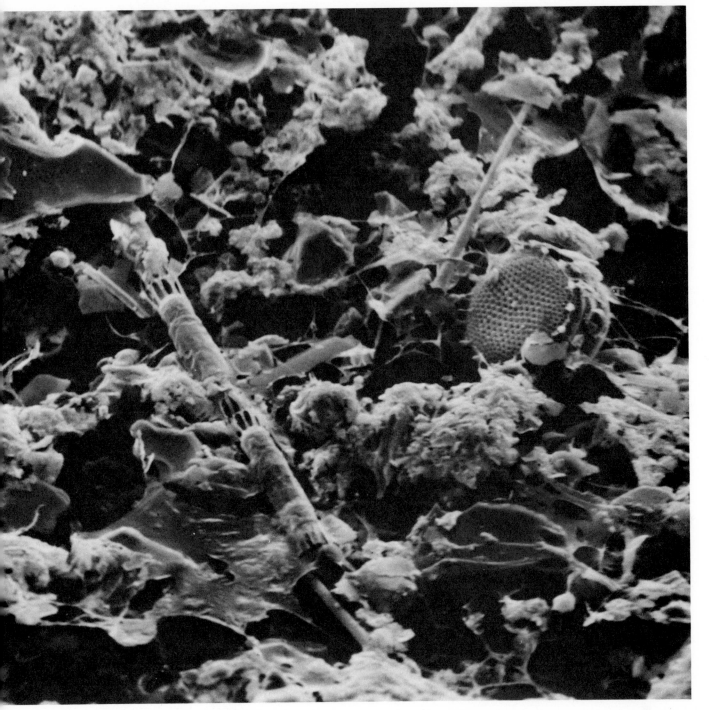

Turbid waters are caused by suspended organic debris, plankton, and mineral grains.

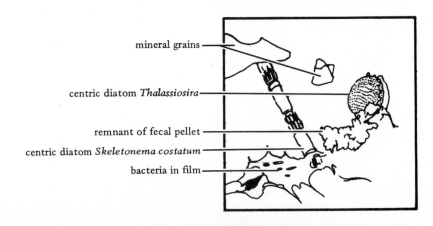

mineral grains

centric diatom *Thalassiosira*

remnant of fecal pellet

centric diatom *Skeletonema costatum*

bacteria in film

Plate 2-3. Particles from the water column concentrated on a filter
Site: Narragansett Bay, RI
Magnification: 1,830X

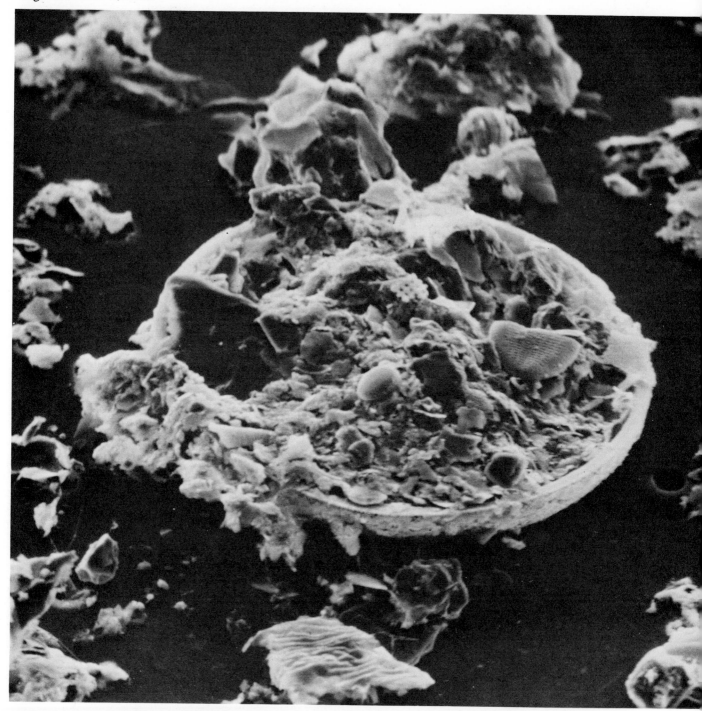

The range in size of diatoms is indicated by the small diatom and fragments inside a valve of a larger diatom.

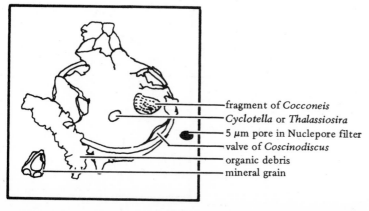

fragment of *Cocconeis*
Cyclotella or *Thalassiosira*
5 µm pore in Nuclepore filter
valve of *Coscinodiscus*
organic debris
mineral grain

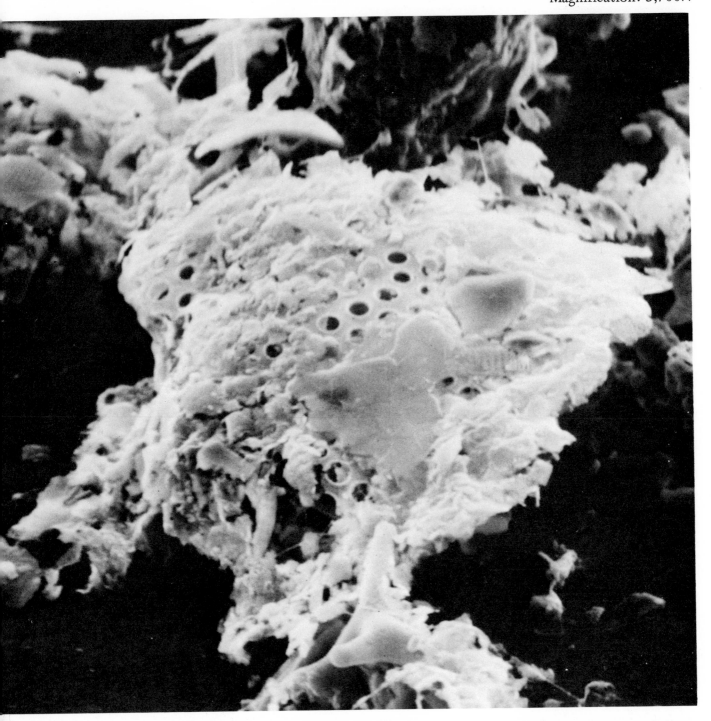

Some of the organic debris are fragments of diatoms encrusted with bacterial slime—

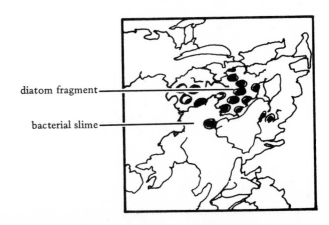

diatom fragment

bacterial slime

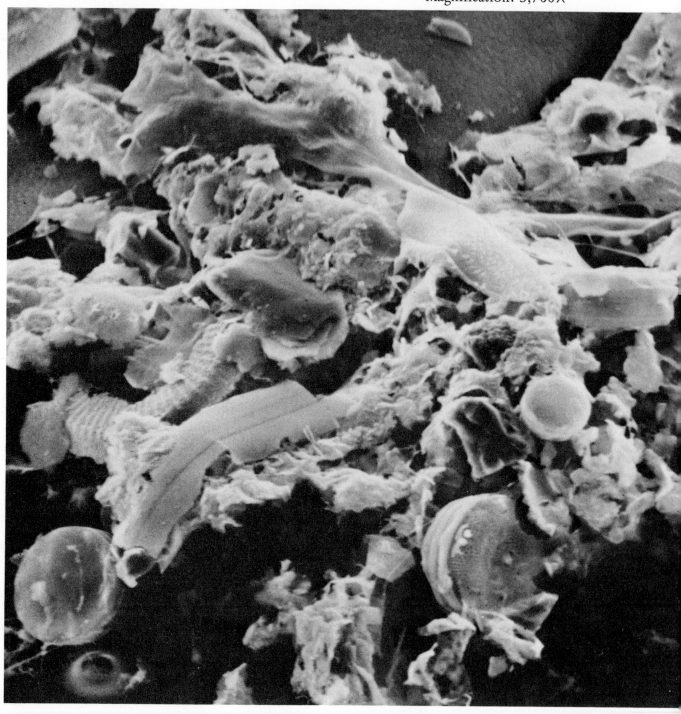

—much of which originates from fragmenting fecal pellets such as this one.

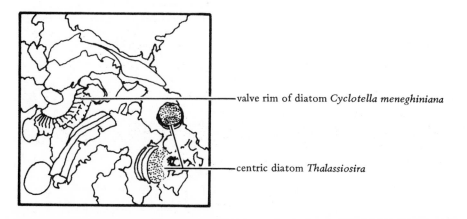

valve rim of diatom *Cyclotella meneghiniana*

centric diatom *Thalassiosira*

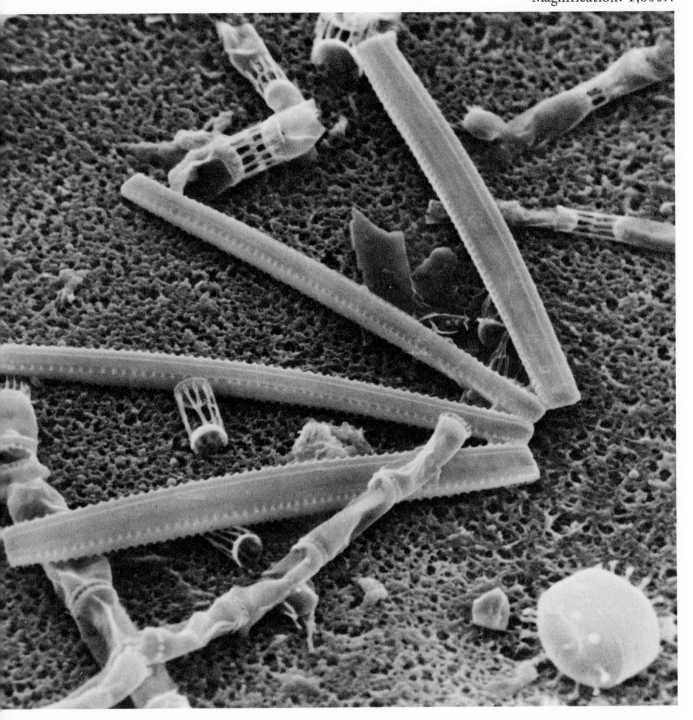

Collecting phytoplankton by net eliminates much of the debris and phytoflagellates but captures the larger diatoms, which are free of attached bacteria—

centric diatom *Skeletonema costatum*

Millipore membrane

pennate diatom *Thalassionema nitzschioides*

centric diatom *Detonula confervacea*

centric diatom *Thalassiosira nordenskjoldii*

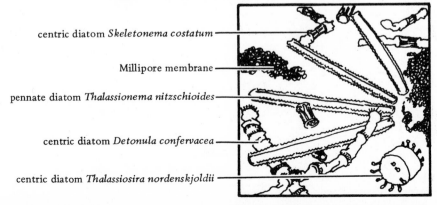

Plate 2-7. Rare bacterial attachment to diatoms
Site: Narragansett Bay, RI
Magnification: 1,780X

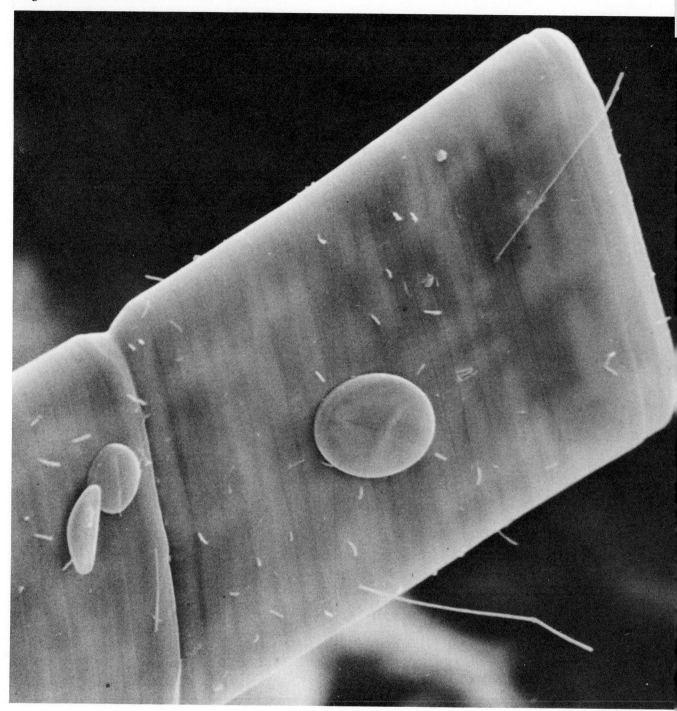

—while on occasion certain species of pennate diatoms have some bacterial colonization.

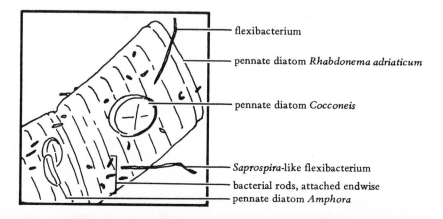

flexibacterium

pennate diatom *Rhabdonema adriaticum*

pennate diatom *Cocconeis*

Saprospira-like flexibacterium

bacterial rods, attached endwise

pennate diatom *Amphora*

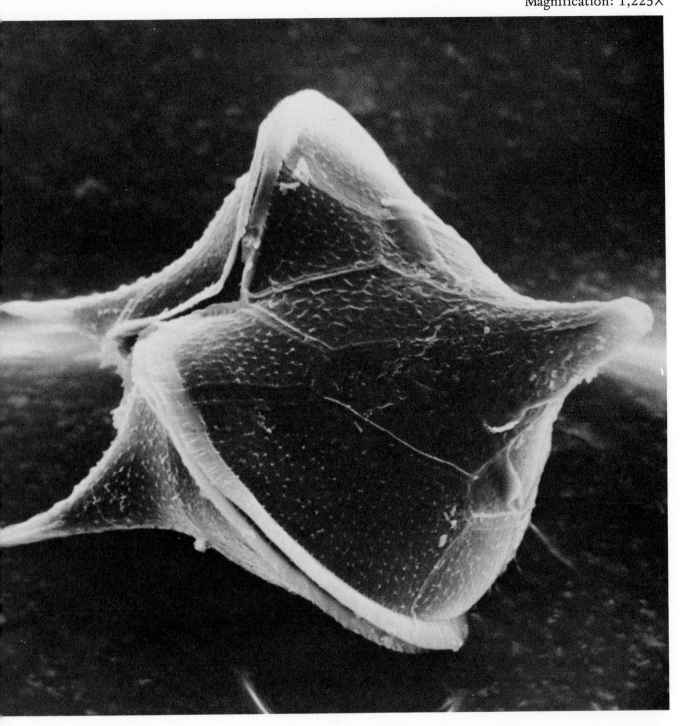

A number of classes of flagellates including this dinoflagellate are important in phytoplankton productivity—

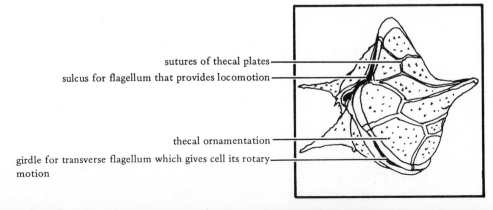

sutures of thecal plates

sulcus for flagellum that provides locomotion

thecal ornamentation

girdle for transverse flagellum which gives cell its rotary motion

Plate 2-9. Dinoflagellate *Gonyaulax tamarensis*
Site: Gulf of Maine
Magnification: 5,300X

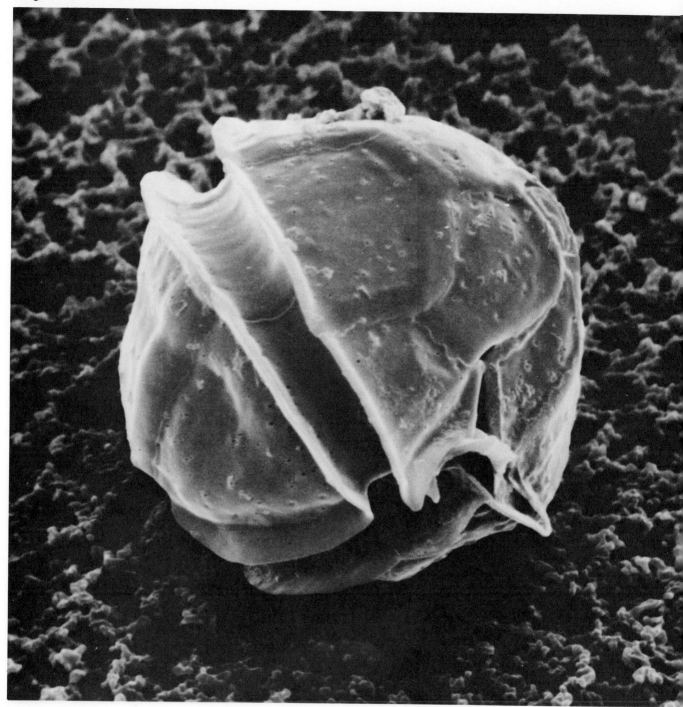

—while a few species under exceptional conditions cause red tides and a paralytic shellfish poisoning of man.

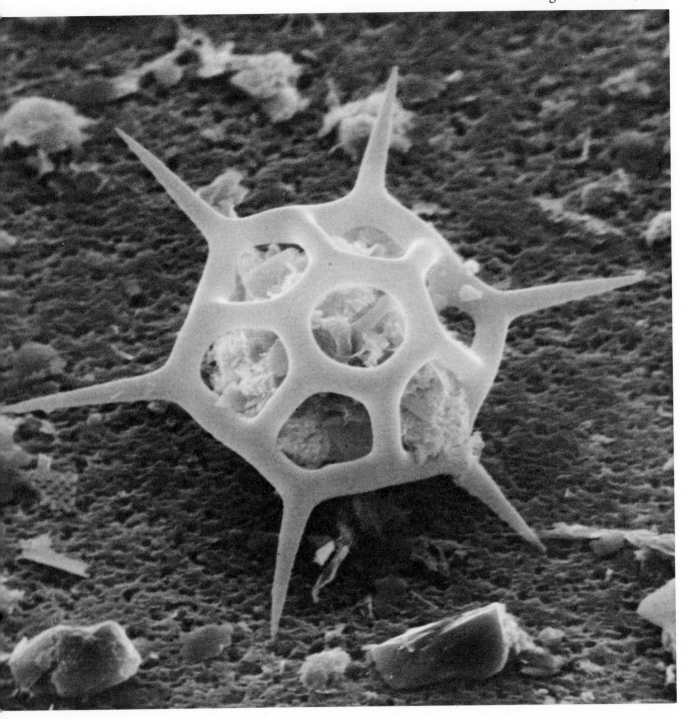

This pleasing and delicate form is a photosynthetic flagellate—

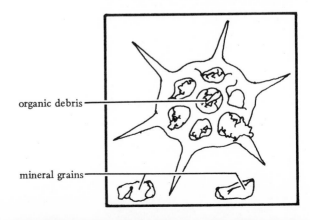

organic debris

mineral grains

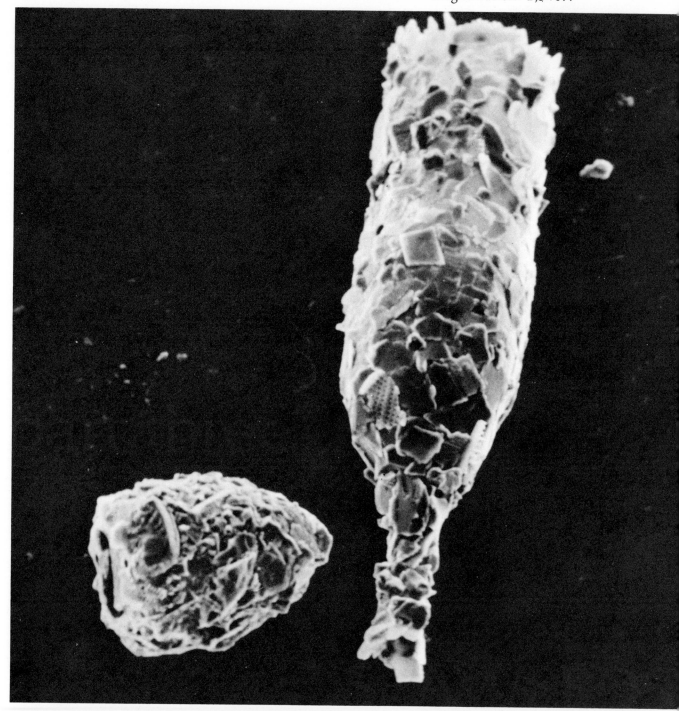

—while these open-ended houses (agglutinated tests) ornamented with debris contain ciliated protozoans that ingest dinoflagellates and other phytoplankters by their ciliary action.

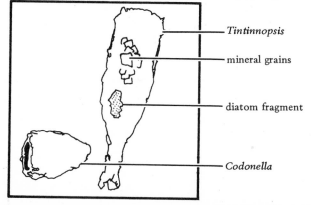

Tintinnopsis

mineral grains

diatom fragment

Codonella

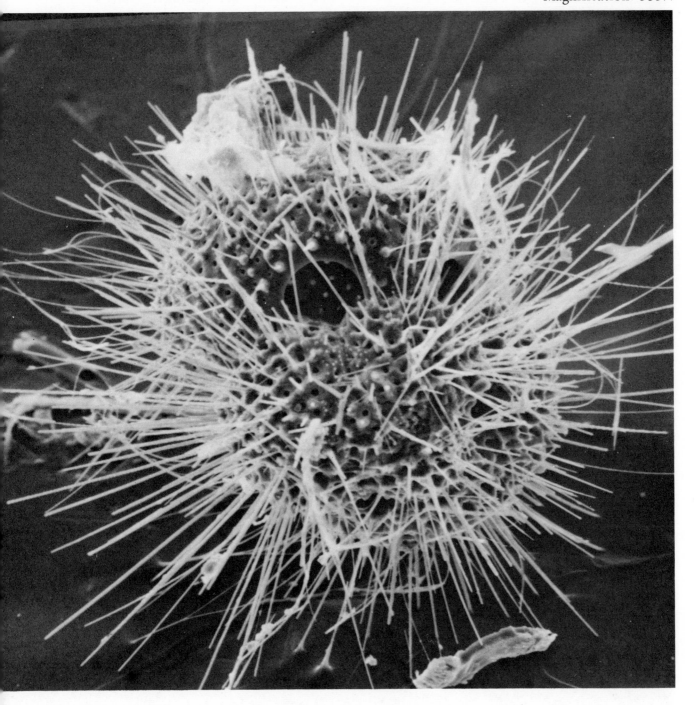

An amoeboid protozoan lives within this multichambered calcareous house. It lives on planktonic microorganisms captured by sticky pseudopods extended through the apertures—

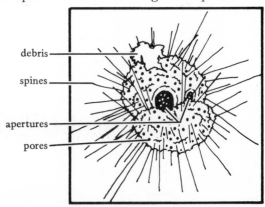

Plate 2-13. Surface of foraminifer
Site: Pigeon Key, FL
Magnification: 5,280X

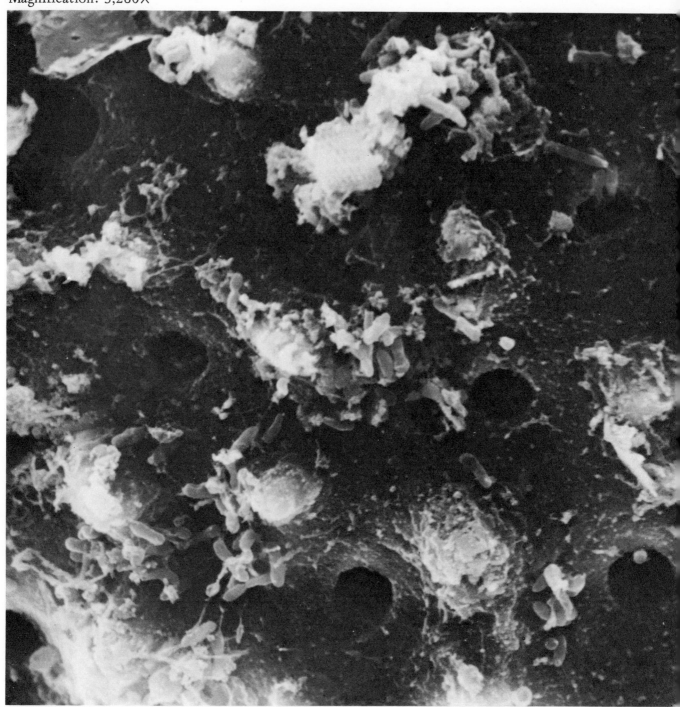

—while its byproducts are excreted through pores in its calcareous test.

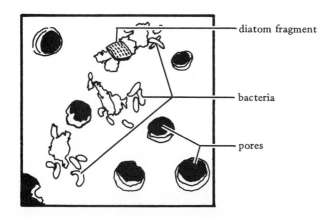

part three

SEAWEEDS

Part of the seaweed forest at the rim of the sea.

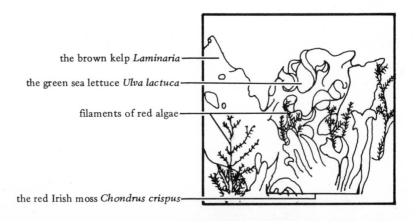

the brown kelp *Laminaria* —

the green sea lettuce *Ulva lactuca* —

filaments of red algae —

the red Irish moss *Chondrus crispus* —

seaweeds

The phytoplankton, although living a buoyant and unattached life, have been called the grass of the sea, upon which the herbivorous species of the zooplankton graze. Similarly, the larger multicellular algae, called seaweeds, may be thought of as aquatic shrubs and trees. Seaweeds occur wherever there are solid substrates like rocks, pilings, and mooring lines to attach to, and wherever light can penetrate at useful intensities.

The seaweeds are divided into the brown algae (*Phaeophyta*), the red algae (*Rhodophyta*), and the green algae (*Chlorophyta*). Each of these divisions contains a great variety of forms, ranging from those with fine filaments and blades in the reds and greens, to those with giant stalk-like stipes and ribbon-like blades longer than 30 meters in the browns. The larger algae can form tremendous standing crops of living biomass in temperate and polar waters. Unlike trees, the seaweeds' rate of growth can be very rapid. The blades of *Laminaria* species grow at the meristem (lower end near the stipe) and decay at the upper (distal) end, producing a number of standing crops each growing season (Mann, 1973). Like trees in the terrestrial environment, seaweeds provide protected nesting grounds for many species of animals (such as fish), food for other animals (such as sea urchins, fish, and snails), and living surfaces for other seaweeds, invertebrate animals, and many types of microorganisms. This is apparent in some of our micrographs.

The best known seaweeds are a few species which have commercial uses. General information on this subject is contained in Tressler's text (1951) on marine products of commerce. Some species are used as snacks or to flavor food, some are applied to the land as fertilizer, while others are dried and used as feed additives for farm animals. Phycocolloids or polysaccharides, which are extracted from seaweeds for industrial, medicinal, and food applications, include alginate from a number of brown seaweeds, carageenin from the red alga *Chondrus crispus,* and agarose from several red algal species. The chemistry of these and other algal polysaccharides is discussed by Percival and McDowell (1967).

We are concerned here with the seaweed surface as an environment for populations of epibiotic microorganisms. The geographic and seasonal variation of the epibiota on the fabled *Sargassum* weed from the Sargasso Sea (Conover and Sieburth, 1966) was found to be caused by polyphenolic tannins (Sieburth and Conover, 1965), one of a number of antibiotic-like inhibitory substances produced by algae (Sieburth, 1968) which affect not only the microorganisms but the larger epibionts which settle later (Ryland, 1974). Extracts of seaweeds have been added to paints to confer antifouling properties (Sieburth, 1966). This effect, however, is on the degree of infestation rather than resulting in an absolute inhibition, as can be seen in the micrographs. Another important factor in the regulation of the fouling on seaweed surfaces is the proteinaceous cuticle which occurs on some species of seaweeds (Hanic and Craigie, 1969). The patchy nature of some algal fouling and the constant renewal of clean algal surfaces in a number of species seem to be caused by the microzonal sloughing of this cuticle or a similar layer, shown in the micrographs and discussed by Tootle (1974). The seaweed surface therefore regulates its fouling by using inhibitory substances and ablative (separating) surfaces just as the marine paint industry does. Despite these regulatory mechanisms, most seaweeds do provide a substrate for rich epibiotic growth, at least seasonally and microzonally.

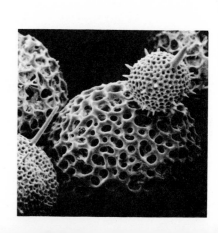

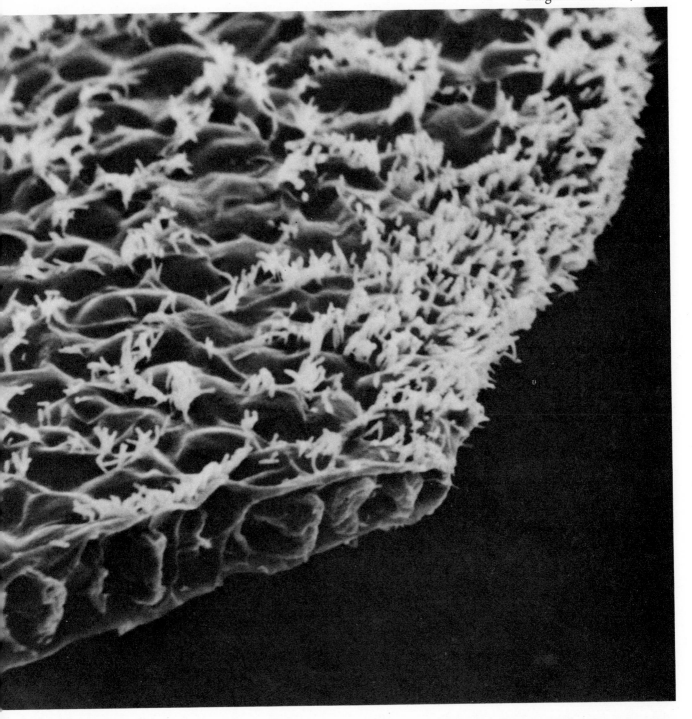

The single layer of algal cells of Monostroma *supports one morphological type of bacteria attached endwise like bristles, that are most dense on the decaying edge—*

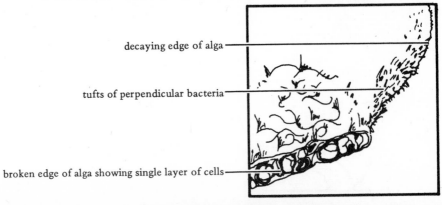

decaying edge of alga

tufts of perpendicular bacteria

broken edge of alga showing single layer of cells

Plate 3-3. The leafy green alga *Ulva lactuca* (sea lettuce)
Site: Narragansett Bay, RI
Magnification: 4,440×

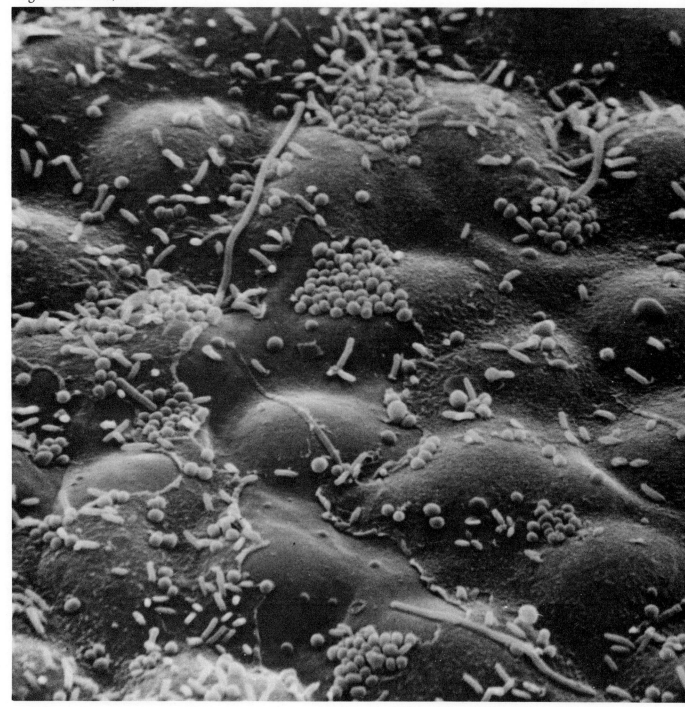

—while the very similar Ulva, *with a double layer of cells, supports a variety of coccoid, rod, and filamentous bacterial forms attached edgewise, but like* Monostroma *is free of diatoms.*

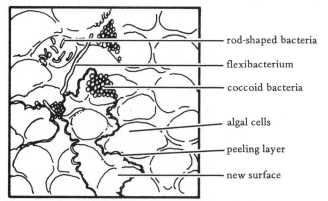

rod-shaped bacteria

flexibacterium

coccoid bacteria

algal cells

peeling layer

new surface

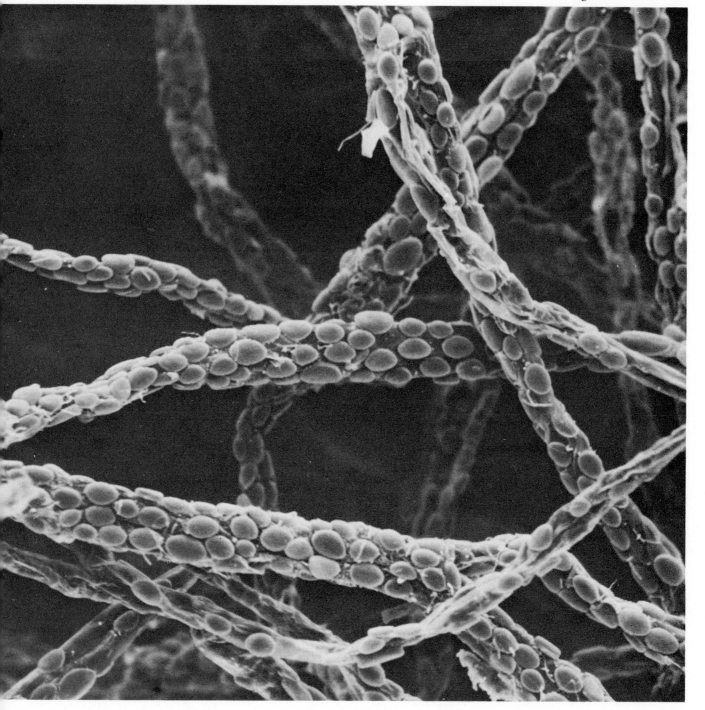

A species of the pennate diatom Cocconeis *has completely covered a* Cladophora *plant—*

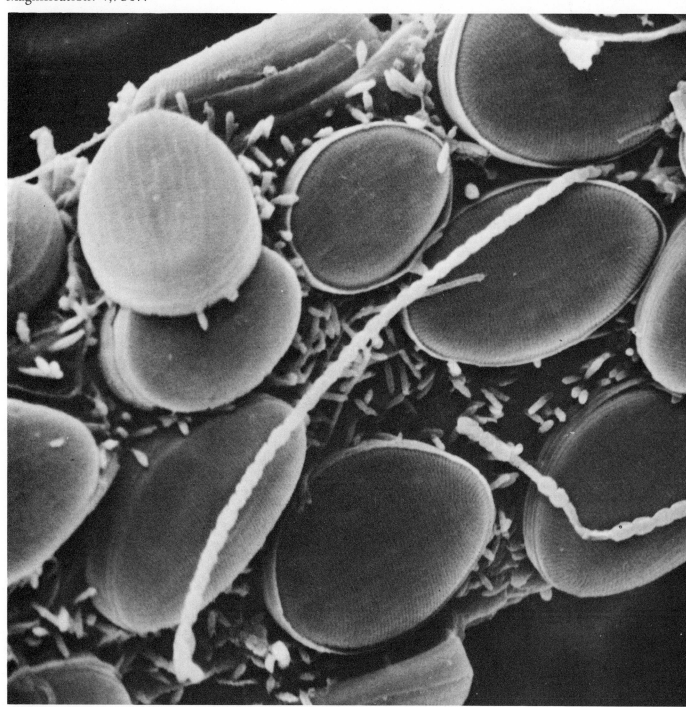

—and is apparently competing with bacteria for algal surface.

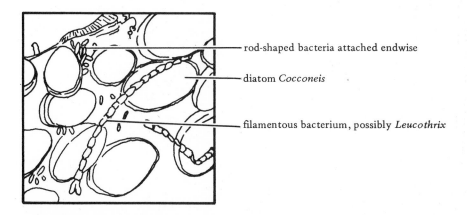

rod-shaped bacteria attached endwise

diatom *Cocconeis*

filamentous bacterium, possibly *Leucothrix*

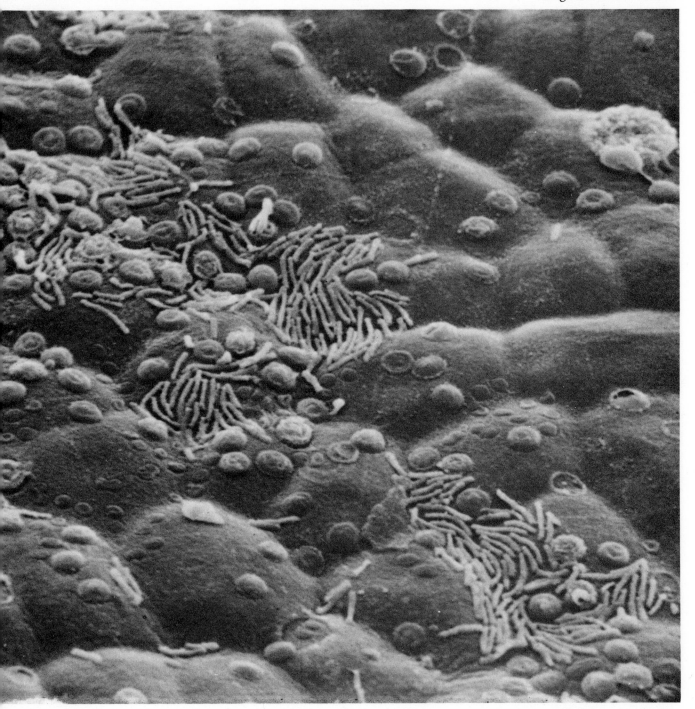

Like the leafy greens, this alga supports a simple bacterial flora—

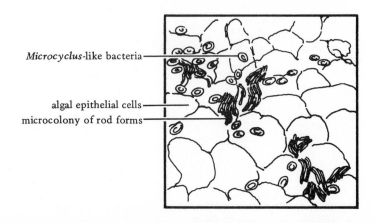

Microcyclus-like bacteria ——

algal epithelial cells ——
microcolony of rod forms ——

Plate 3-7. The leafy red alga *Rhodymenia palmata* (dulse)
Site: Narragansett Bay, RI
Magnification: 3,885×

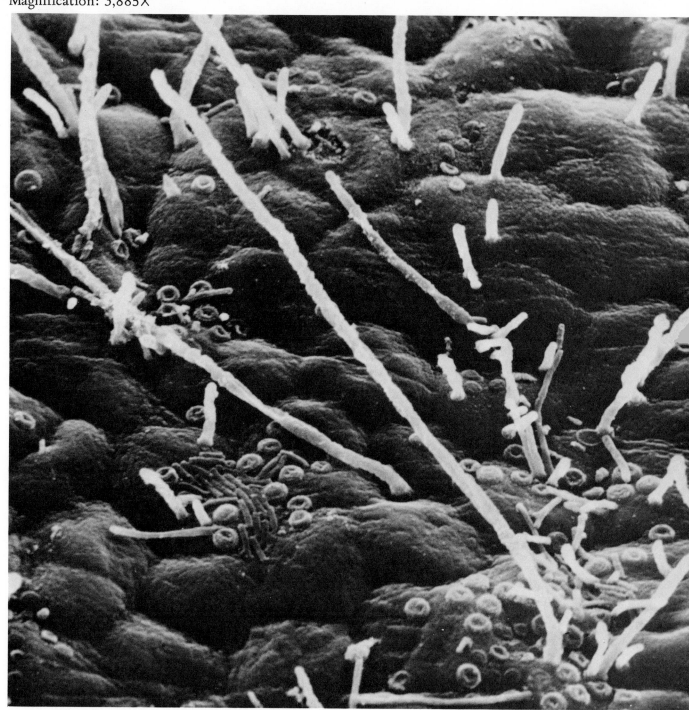

—*including rings, rods, and filaments.*

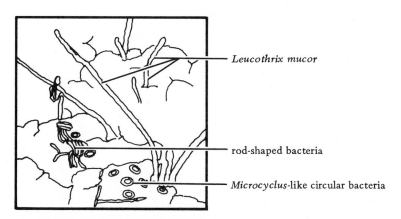

Leucothrix mucor

rod-shaped bacteria

Microcyclus-like circular bacteria

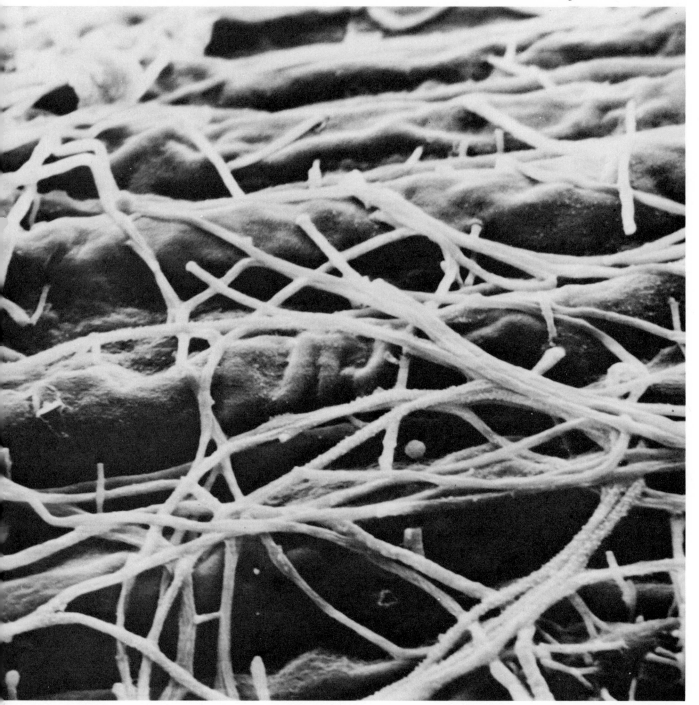

The corrugated filaments of Polysiphonia *are covered year round with bacterial filaments (possibly*
Leucothrix *)—*

Plate 3-9. The filamentous red alga *Polysiphonia lanosa*
Site: Narragansett Bay, RI
Magnification: 4,800×

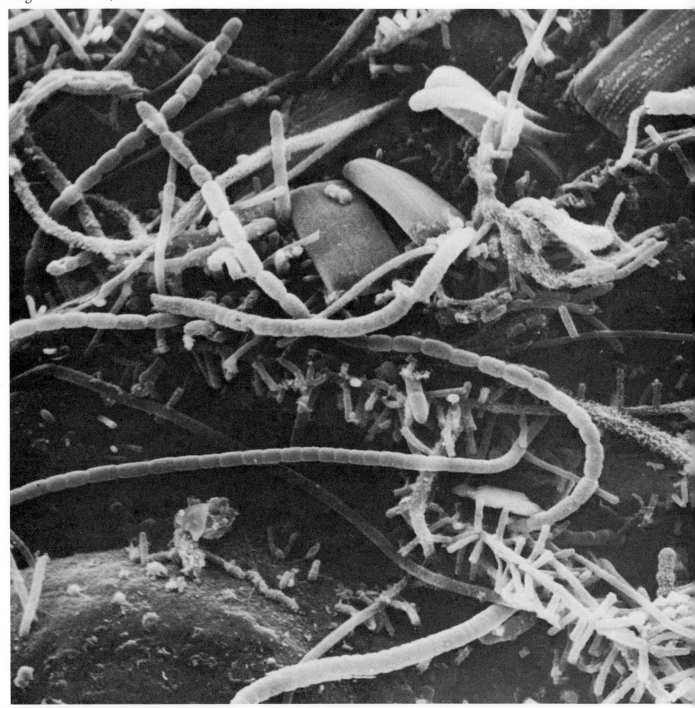

—and occasionally with other filaments and pennate diatoms.

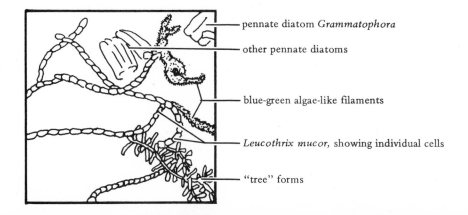

pennate diatom *Grammatophora*

other pennate diatoms

blue-green algae-like filaments

Leucothrix mucor, showing individual cells

"tree" forms

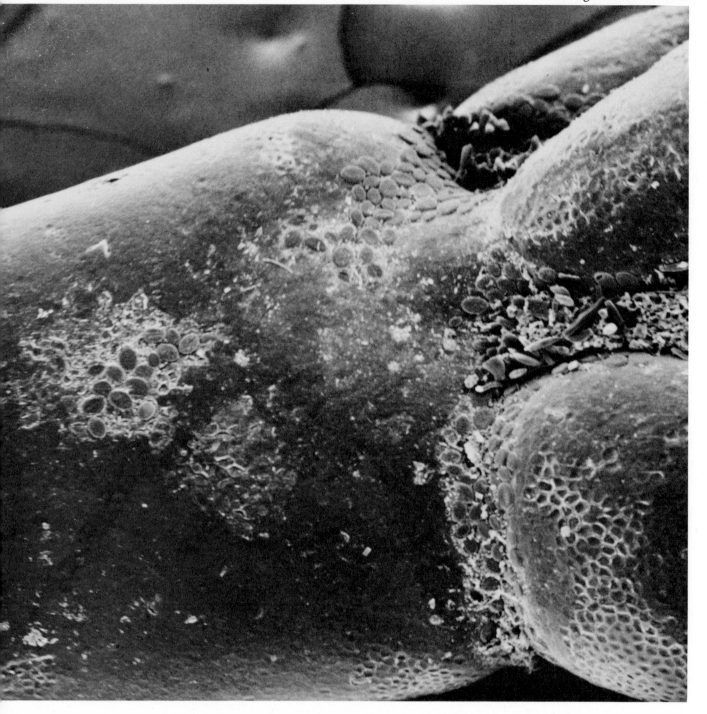

The protected areas and exposed cells with their calcified structures are colonized by diatoms—

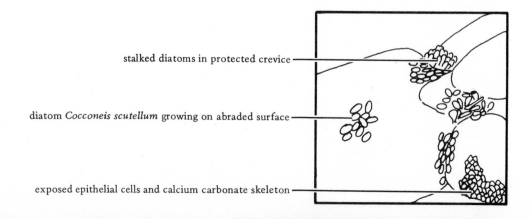

stalked diatoms in protected crevice

diatom *Cocconeis scutellum* growing on abraded surface

exposed epithelial cells and calcium carbonate skeleton

Plate 3-11. Diatoms encrusting coralline alga
Site: Narragansett Bay, RI
Magnification: 4,940X

—*which appear to sink into the algal surface while bacteria develop on exposed areas.*

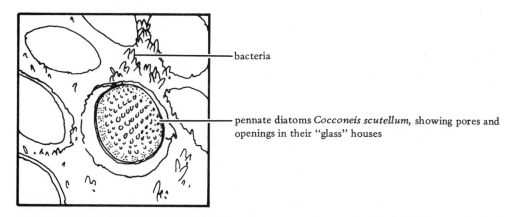

bacteria

pennate diatoms *Cocconeis scutellum,* showing pores and openings in their "glass" houses

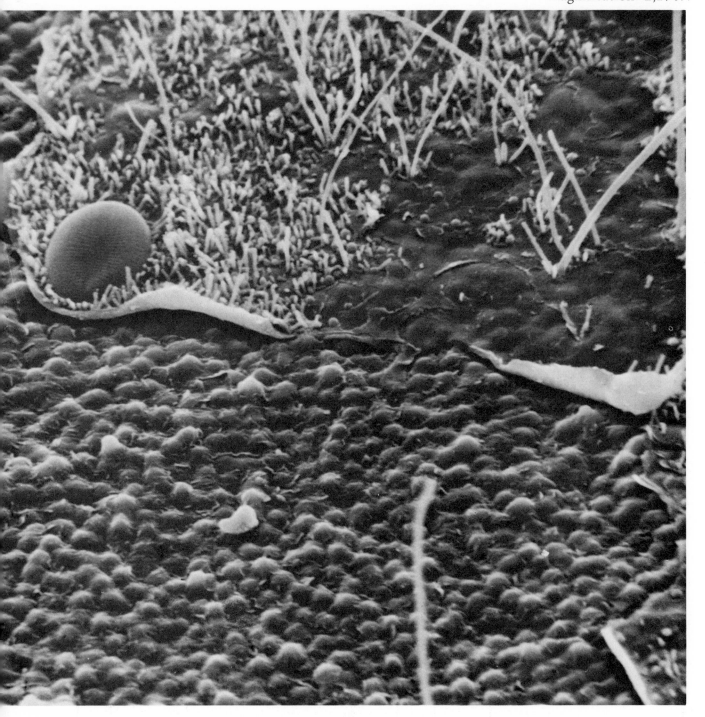

The fouled surface peels off when the cuticle sloughs, to expose a fresh algal surface—

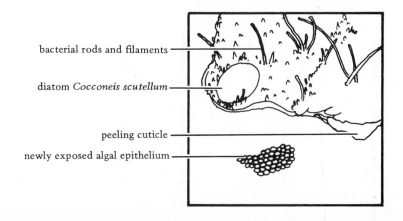

bacterial rods and filaments

diatom *Cocconeis scutellum*

peeling cuticle

newly exposed algal epithelium

Plate 3-13. The leafy brown alga *Laminaria longicruris* (kelp)
Site: Narragansett Bay, RI
Magnification: 1,600X

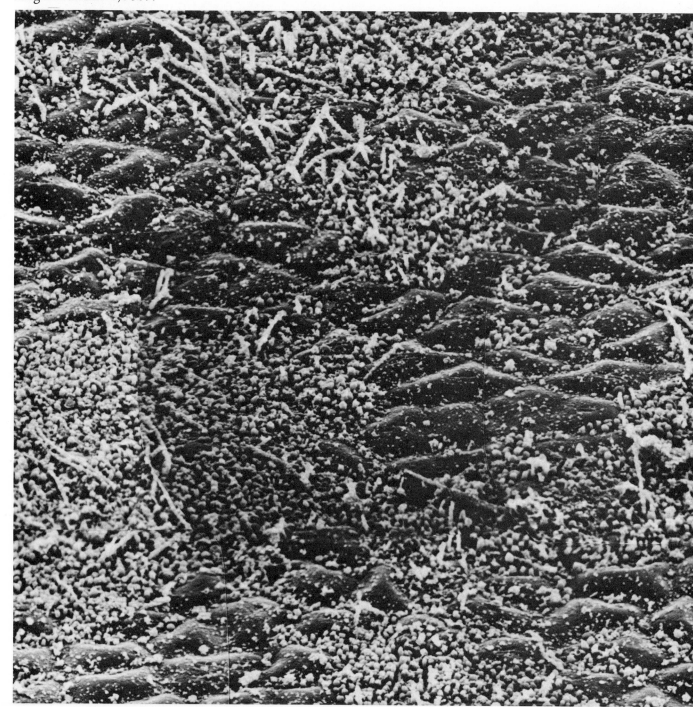

—which is then recolonized, as shown by contrasting areas of older and new fouling.

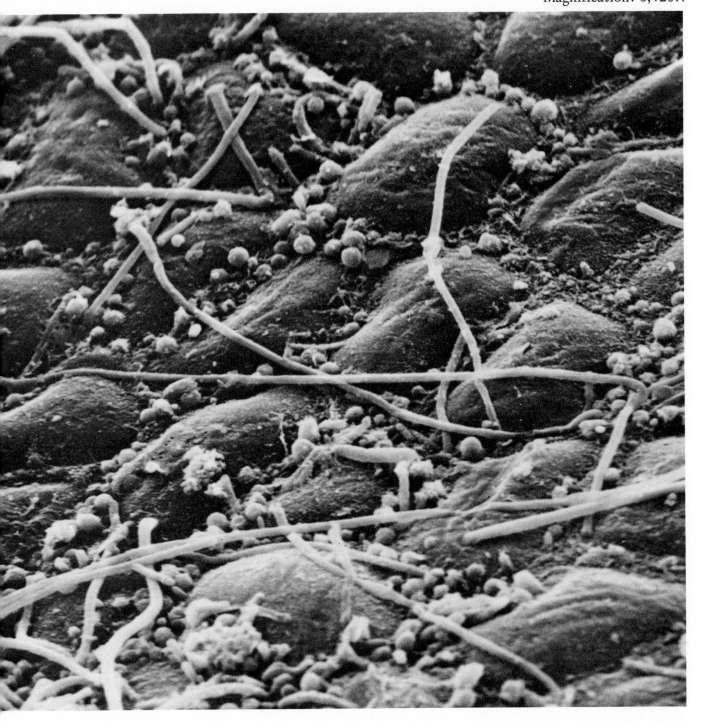

Certain algae release inhibitory substances during their active growing period in spring and are lightly fouled—

Plate 3-15. The brown alga *Ascophyllum nodosum* (knotted wrack)
Site: Narragansett Bay, RI—October
Magnification: 5,370X

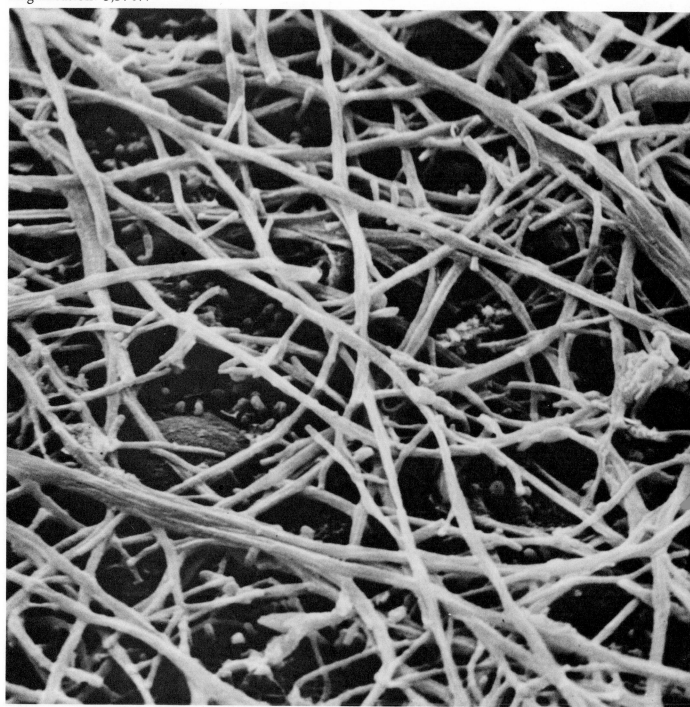

—*but become overgrown with bacterial filaments by autumn.*

Plate 3-16. Leafy blade of the brown alga *Sargassum natans* (gulfweed)
Site: Pigeon Key, FL
Magnification: 80X

Tufts of blue-green algae—

Plate 3-17. Surface of *Sargassum natans*
Site: Pigeon Key, FL
Magnification: 2,570X

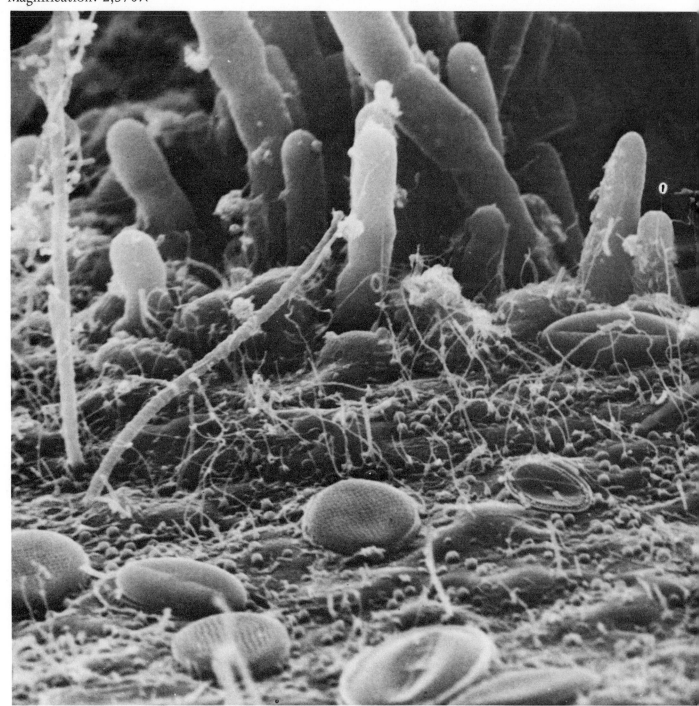

—arise from a surface that can support a rich microbiota when the production of inhibitory polyphenols is minimized by seasonal and geographic factors.

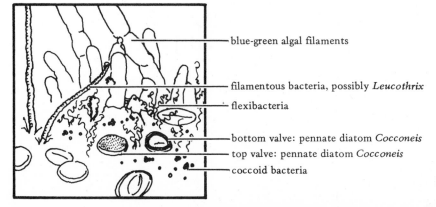

blue-green algal filaments

filamentous bacteria, possibly *Leucothrix*

flexibacteria

bottom valve: pennate diatom *Cocconeis*
top valve: pennate diatom *Cocconeis*
coccoid bacteria

part four

SEA GRASSES

The plants and animals of a Thalassia community are dusted with the calcareous sand of this tropical environment.

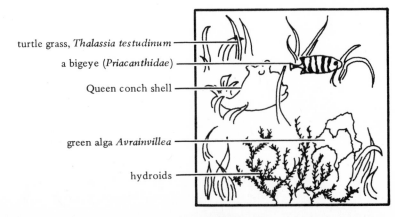

turtle grass, *Thalassia testudinum*

a bigeye (*Priacanthidae*)

Queen conch shell

green alga *Avrainvillea*

hydroids

sea grasses

The seaweeds, which require rocky coasts and turbulent waters, give way to sea grasses in protected shallow waters with muddy bottoms, such as salt ponds, estuaries, and reef platforms. The sea grasses, their debris, and the microorganisms which develop upon both, are very important to the ecology of these areas. The grass beds, like the seaweed beds, also provide a nursery for many species of vertebrates and invertebrates. The early stages of numerous fish—including snappers, pompano, barracuda, and mullet—and crustacea, such as the spiny lobster and many species of shrimp, thrive in the southern grass beds, while flounders, shrimp, and crabs, among others, abound in the grass beds of temperate climates.

The microbial epibionts that develop on the standing grasses support a variety of larger forms including foraminifera (amoebae encased in calcareous tests), nematodes (microscopic worms), tube-dwelling animals, and a variety of snails which forage over the fouled grass surface.

The grasses are eaten directly by ducks, geese, and other waterfowl in the north and by sea urchins, manatees, and parrot fish in the south. The bulk of the grasses, however, break off, fragment, and decompose. The microbial forms responsible for this decomposition enrich the organic debris and form nutrients for a great variety of deposit feeders in the mud (Fenchel, 1970, 1972).

Eelgrass, *Zostera marina,* is readily fouled by microorganisms and larger forms. The blades of these grasses become encrusted and discolored with diatoms, and look fuzzy because they are covered with filamentous algal forms. This sea grass of temperate water can undergo a unialgal primary fouling by the pennate diatom *Cocconeis scutellum,* followed by non-selective colonization to form a thick microbial crust (Sieburth and Thomas, 1973). The animal colonization of eelgrass has been described by Marsh (1970). The apparent paucity of crawling protozoans and animal forms in most of our micrographs may be attributable to their exclusion by our preparatory techniques.

In the waters from south Florida and along the coast of the Gulf of Mexico, eelgrass is replaced by extensive stands of turtle grass, *Thalassia testudinum.* The rhizomes of this grass extend down to 50 cm into the sediment and stabilize the bottom sediments, which are formed in part by the blades of the grass acting as a sediment trap. In addition to the dominant turtle grass with its flat blades 12 mm wide, shoal grass (*Diplantheria wrightii,* with flat blades 2 mm wide) and manatee grass (*Syringodium filiforme,* with its tubular blade) are also common in their own beds or mixed with the turtle grass (Zischke, 1973). These southern sea grasses can also be colonized by a species of the pennate diatom *Cocconeis,* but bacteria seem to play a greater role here than in the colonization of eelgrass.

A variety of marsh and beach grasses grow higher up in the tidal zone. The ecology of these grasses in the salt marsh has been discussed by Jefferies (1972). Most of these grasses grow high and dry with only their lower portion immersed by the rising tides. The fungal microflora developing on the immersed portion of the cord grass, *Spartina alterniflora,* was described by Gessner, Goos, and Sieburth (1972). When the marsh and beach grasses get knocked down in the fall, they find their way into the shallow brackish water and, like the sea grasses, form organic debris and a microbial flora and fauna which supply food to deposit feeders such as the grass shrimp *Palaemonetes pugio* (Welsh, 1973).

The micrographs used in this section span a 3-year period of our observations. If one compares the micrographs of eelgrass made with freeze-dried preparations to those of turtle grass which were critical point dried, the importance of proper specimen preparation for microscopy becomes apparent. The resolution of the fine architecture of the diatom frustule is vastly improved, and the method of attachment used by the bacteria (endwise or edgewise) is very clearly revealed. Bacteria attached edgewise are in greater contact with the surface upon which their enzymes are acting and releasing soluble compounds. The endwise-attached bacteria may be less dependent upon the surface for nutrients or they may be trying to minimize the effects of an inhibitory surface (Sieburth, 1968).

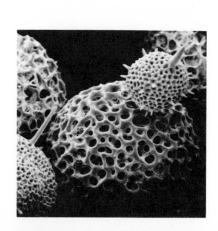

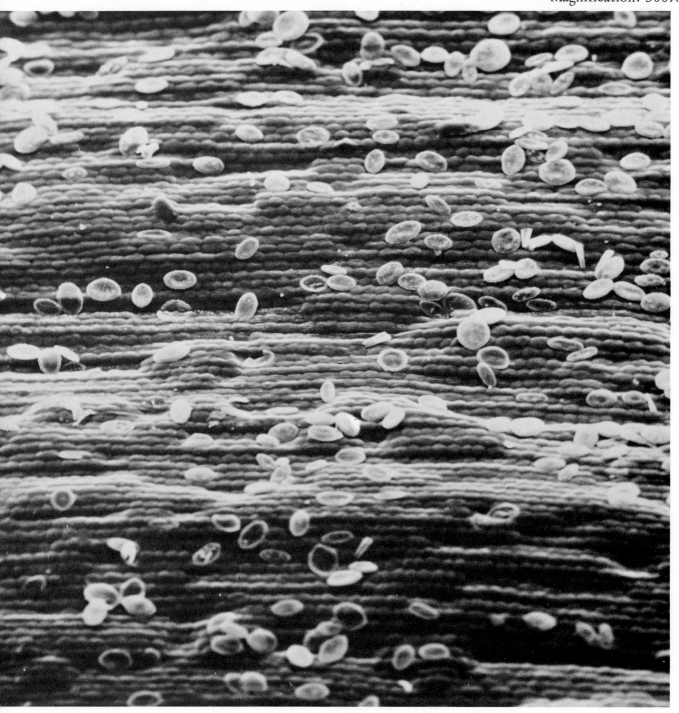

The cobblestone-like surface of the grass is colonized by diatoms—

Plate 4-3. Diatom-covered eelgrass
Site: Pettaquamscutt River, RI
Magnification: 540X

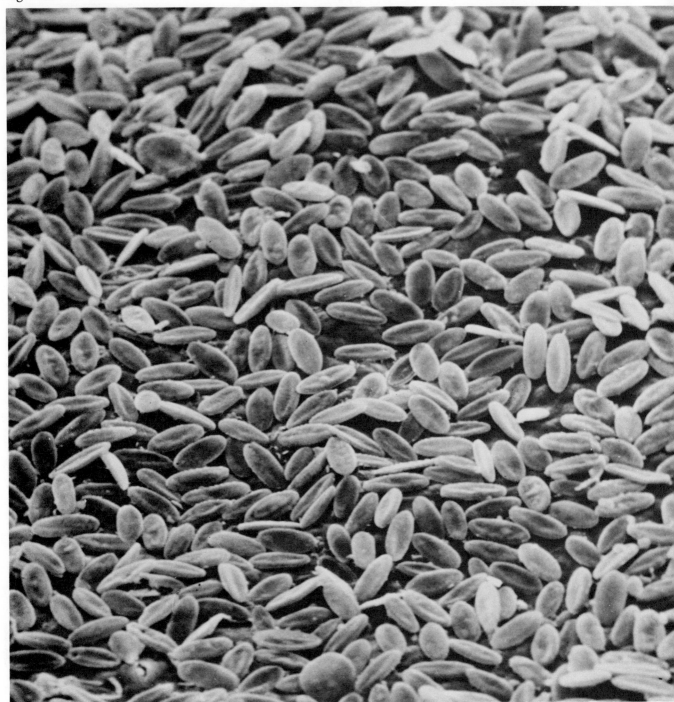

—which multiply to form a single-species layer of Cocconeis scutellum*—*

—that permits a multispecies diatom crust to form—

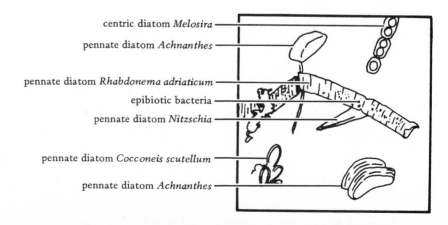

centric diatom *Melosira*

pennate diatom *Achnanthes*

pennate diatom *Rhabdonema adriaticum*

epibiotic bacteria

pennate diatom *Nitzschia*

pennate diatom *Cocconeis scutellum*

pennate diatom *Achnanthes*

Plate 4-5. Peeling diatom crust on eelgrass
Site: Pettaquamscutt River, RI
Magnification: 220×

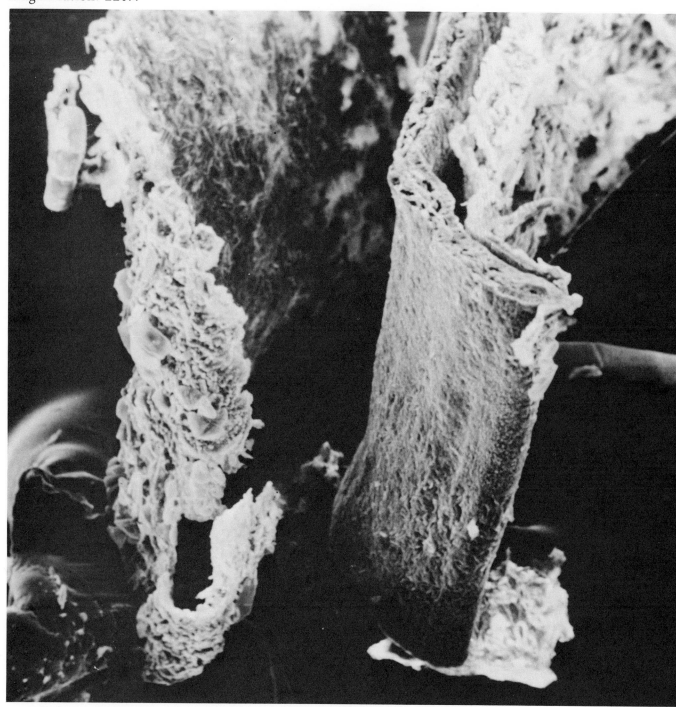

—which can equal the biomass of the host plant.

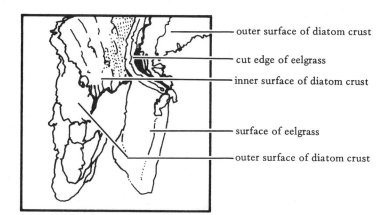

outer surface of diatom crust

cut edge of eelgrass

inner surface of diatom crust

surface of eelgrass

outer surface of diatom crust

Diatom encrustation is not so pronounced on turtle grass—

Plate 4-7. Diatom-bacterial population on turtle grass
Site: Pigeon Key, FL
Magnification: 5,220X

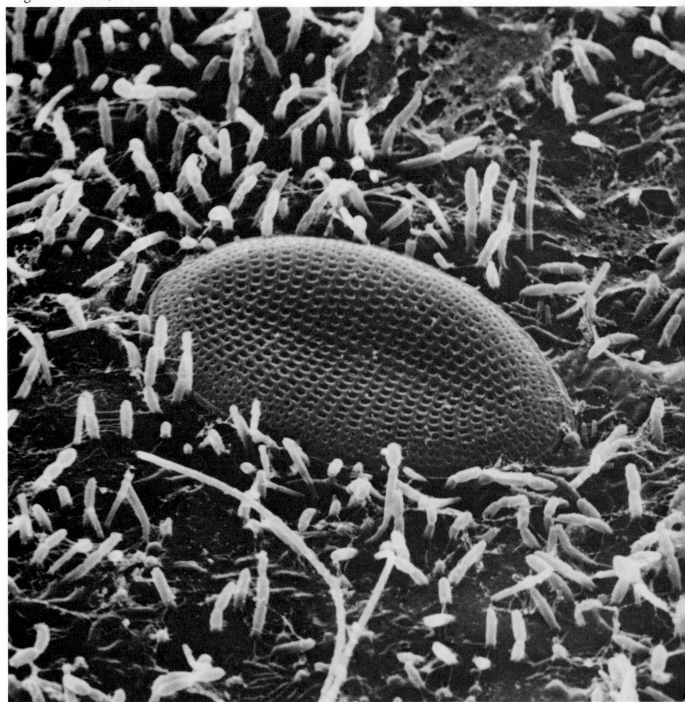

—while bacterial development seems more substantial.

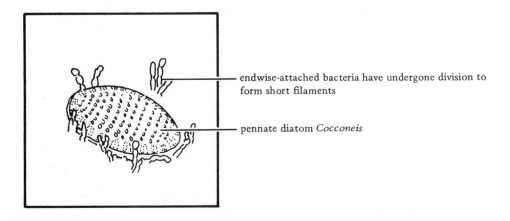

endwise-attached bacteria have undergone division to form short filaments

pennate diatom *Cocconeis*

Some areas are colonized by algal filaments and stalked diatoms visible to the naked eye—

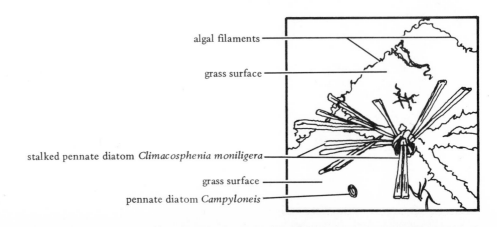

algal filaments

grass surface

stalked pennate diatom *Climacosphenia moniligera*

grass surface

pennate diatom *Campyloneis*

—while the grass surface has a few scattered diatoms and debris.

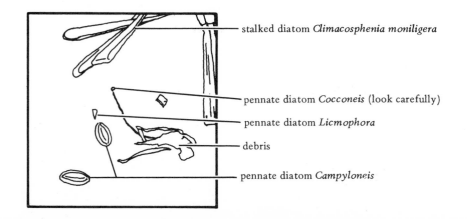

stalked diatom *Climacosphenia moniligera*

pennate diatom *Cocconeis* (look carefully)

pennate diatom *Licmophora*

debris

pennate diatom *Campyloneis*

Plate 4-10. Turtle grass *Thalassia testudinum*
Site: Pigeon Key, FL
Magnification: 1,260X

At magnification sufficient to see clearly the smaller diatoms, an extensive bacterial lawn—

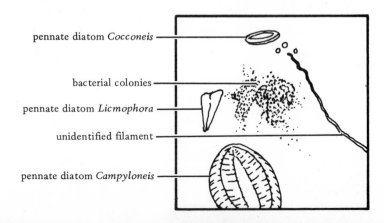

pennate diatom *Cocconeis*

bacterial colonies

pennate diatom *Licmophora*

unidentified filament

pennate diatom *Campyloneis*

Plate 4-11. Turtle grass *Thalassia testudinum*
Site: Pigeon Key, FL
Magnification: 4,770X

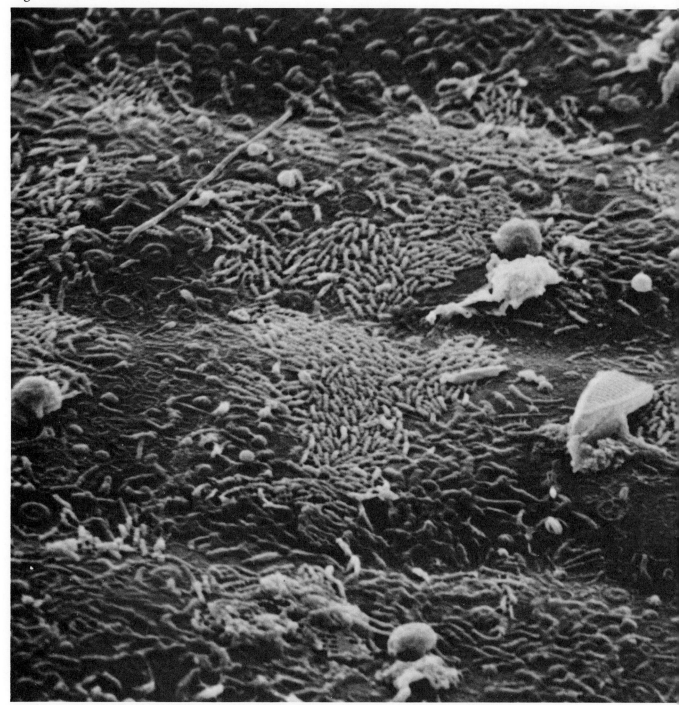

—is seen to be composed of similarly oriented bacterial cells and an occasional distinctive form.

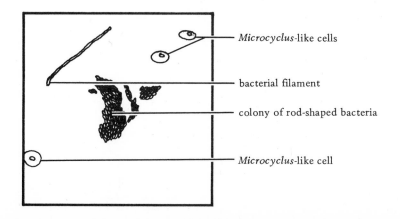

Microcyclus-like cells

bacterial filament

colony of rod-shaped bacteria

Microcyclus-like cell

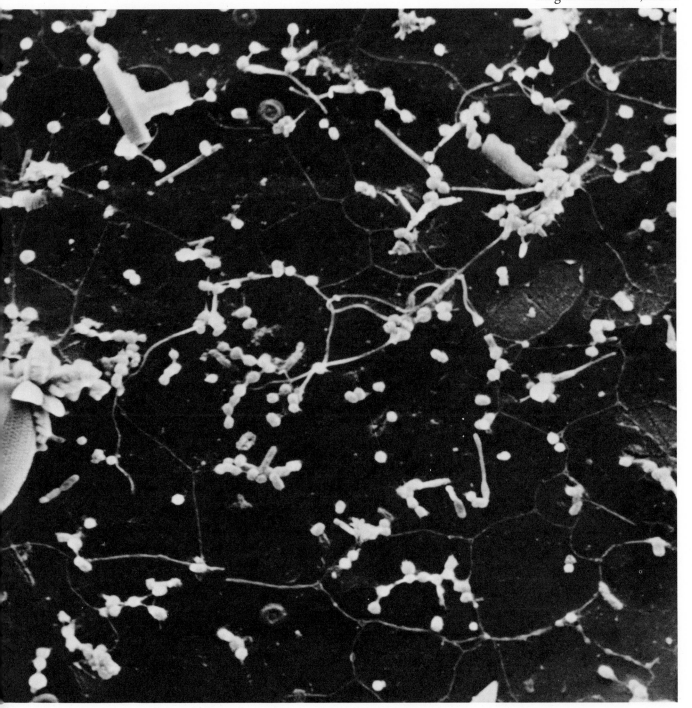

The smooth surface of manatee grass contains a different assortment of bacteria and an occasional diatom—

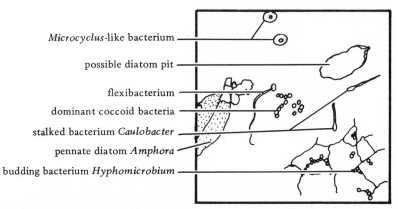

Microcyclus-like bacterium ——

possible diatom pit ——

flexibacterium ——
dominant coccoid bacteria ——

stalked bacterium *Caulobacter* ——
pennate diatom *Amphora* ——

budding bacterium *Hyphomicrobium* ——

Plate 4-13. Blade of manatee grass *Syringodium filiforme*
Site: Pigeon Key, FL
Magnification: 3,400×

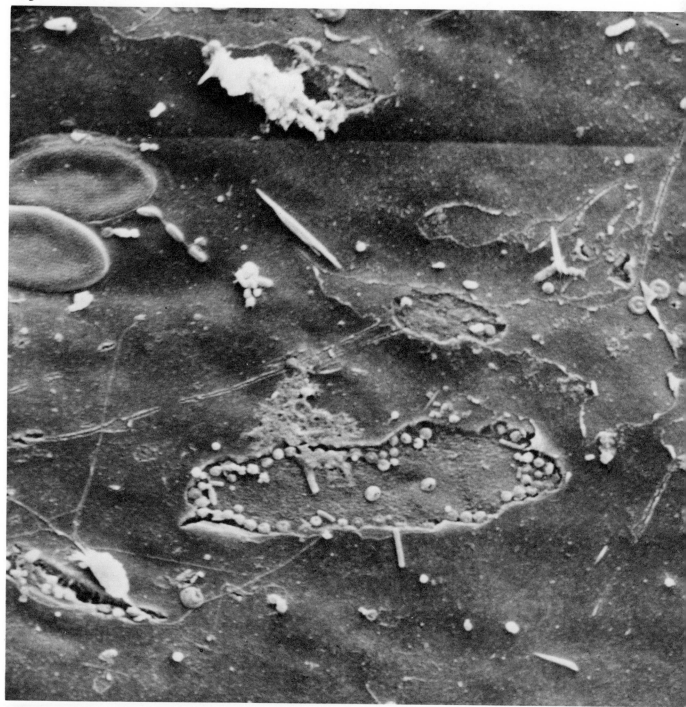

—which seems to penetrate into the surface, leaving pits in which bacteria develop.

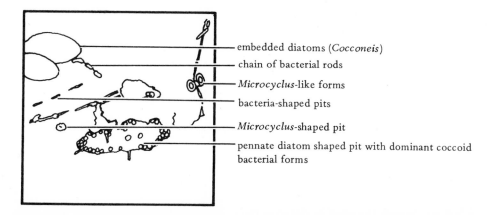

embedded diatoms (*Cocconeis*)
chain of bacterial rods
Microcyclus-like forms
bacteria-shaped pits
Microcyclus-shaped pit
pennate diatom shaped pit with dominant coccoid bacterial forms

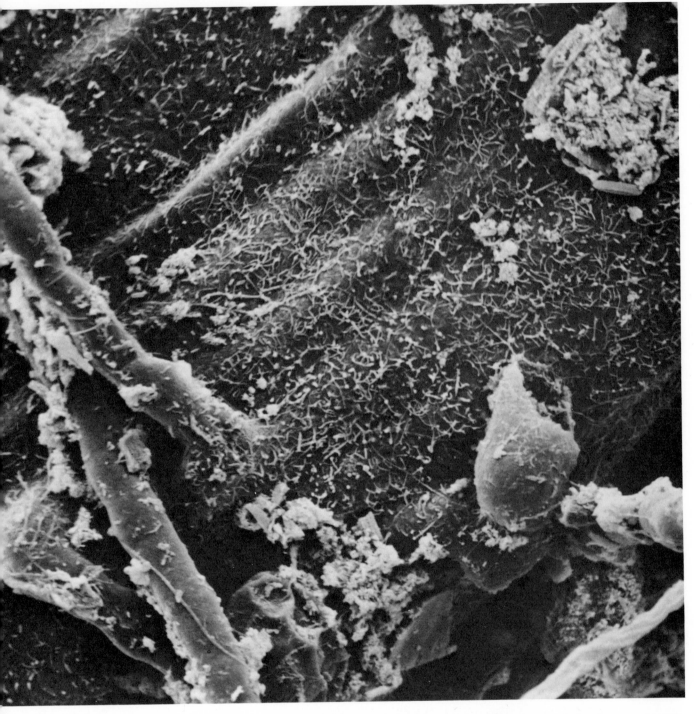

The root hairs seem free of bacteria, while the root itself—

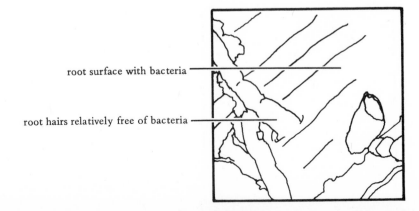

root surface with bacteria ──────────

root hairs relatively free of bacteria ──────────

Plate 4-15. Root of manatee grass *Syringodium filiforme*
Site: Pigeon Key, FL
Magnification: 4,840X

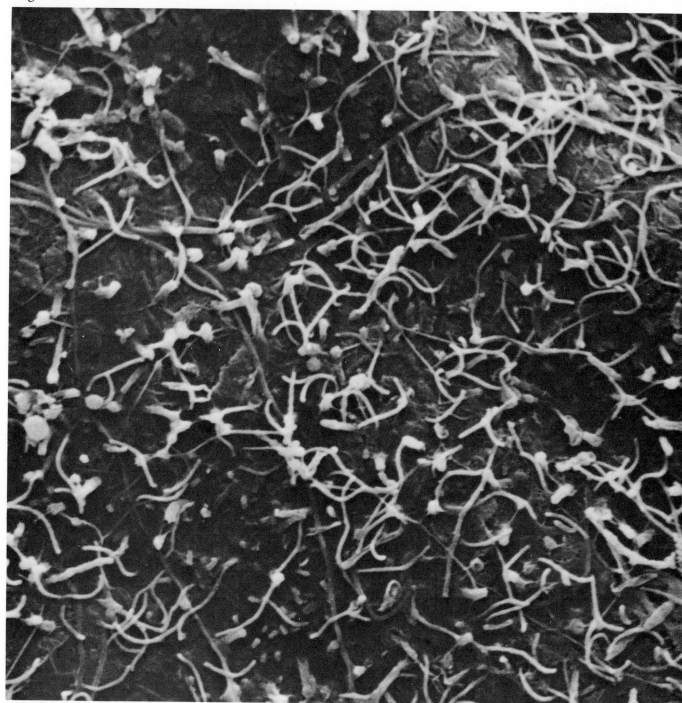

—is heavily colonized by bacteria dominated by filamentous forms.

part five

MAN-MADE SURFACES

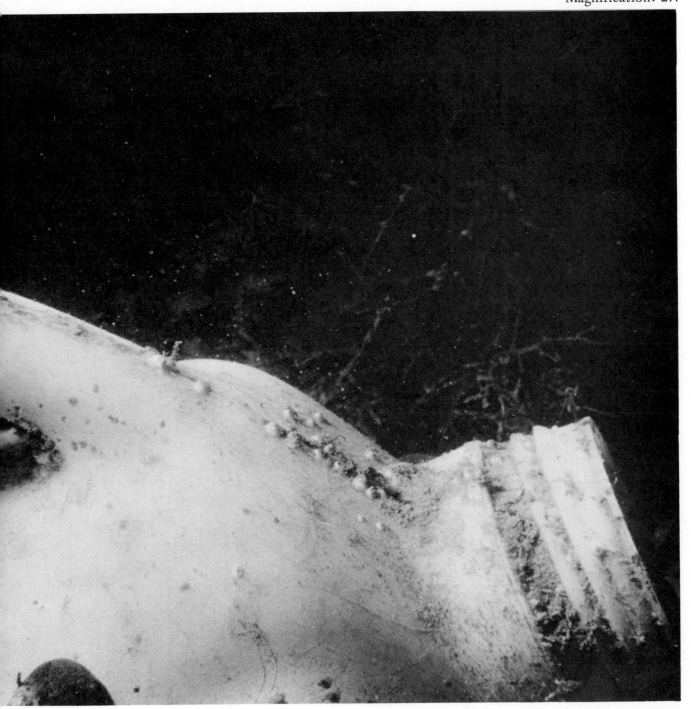

This example of man's refuse is resistant to biodeterioration but becomes involved in the biological cycle of the sea, supporting both patches of diatoms and the settlement of barnacles.

man-made surfaces

A seemingly endless variety of non-living or inanimate surfaces occur in the sea. Organic materials undergoing deterioration and the rocks and mineral debris on the sea floor are covered in Parts Seven and Eight. Most of the other non-living surfaces that foul in the sea are man-made objects such as boats, pilings, buoys, flotsam, and jetsam. The more common term "inanimate" is a poor choice here because, due to microbial colonization, they do not remain inanimate surfaces for long. Most studies on the microbial fouling of such surfaces have been conducted on glass microscope slides because of their transparency and the ease of examination by the light microscope. A good example of such a study (which gives reference to the earlier work) is that by Skerman (1956). Unfortunately, the relatively smooth glass surface which has been used so extensively to study microbial fouling is very unlike the rough, pitted, and cracked surfaces of most other materials and is slower to colonize. Glass can show a very different colonization than an adjacent plastic surface. Metal surfaces are difficult to examine and interpret because of the products of corrosion and oxidation that occur rapidly in sea water.

The typical pattern of microbial fouling on a non-nutrient surface seems to be pennate diatoms or algal filaments as the primary colonizers. They produce the materials necessary to sustain bacteria on the non-nutrient surface. Protozoa and invertebrates such as bryozoans, hydroids, and worms then settle on these microbial films. A bleach bottle trapped subtidally is an excellent example of a diatom, bacterial, invertebrate succession. The varnished wood trim and a piece of the fiberglass hull of a derelict boat recovered in the Atlantic Ocean, north of Bermuda, depicts an algal, bacterial, hydroid succession.

Inshore waters, especially polluted waters, support a rich bacterial flora that can attach itself to immersed surfaces immediately and produce a heavily colonized surface within days. Micrographs of glass, plastic, and a nylon fishing line are used to show this bacteria-first phenomenon.

The plastic surface supports a layer of mixed diatoms—

Plate 5-3. Bacterial colonization of bleach bottle
Site: Narragansett Bay, RI
Magnification: 5,200×

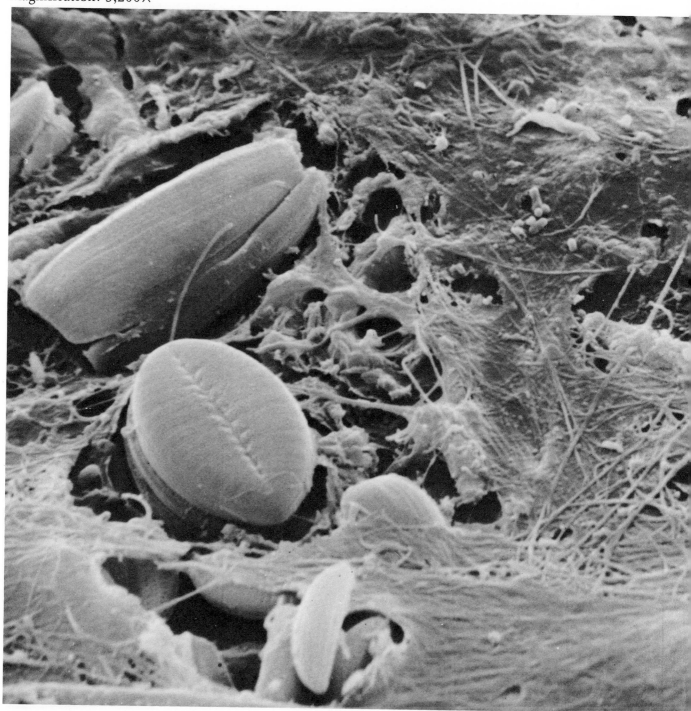

—*which supports a bacterial film.*

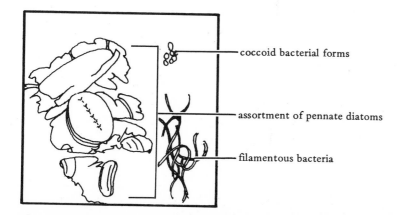

coccoid bacterial forms

assortment of pennate diatoms

filamentous bacteria

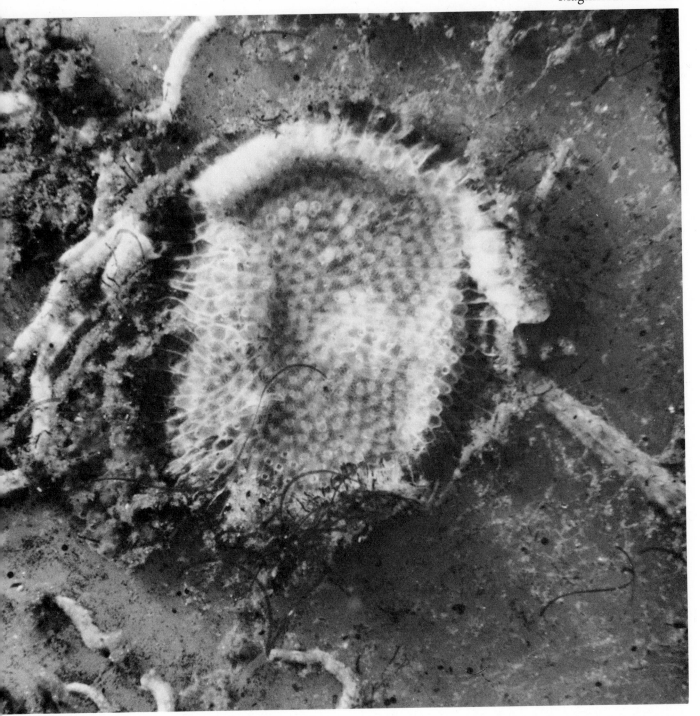

This microbial film seems to be a prerequisite for the settlement of invertebrate forms seen in this macrograph—

tubes of polychaete worm ⎯

blue-green algae ⎯

colony of ascophoran bryozoan ⎯

ribbons of the pennate diatom *Rhabdonema* ⎯

patch of smaller diatoms ⎯

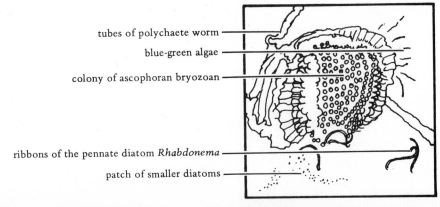

Plate 5-5. Invertebrate colonization of bleach bottle
Site: Narragansett Bay, RI
Magnification: 97 X

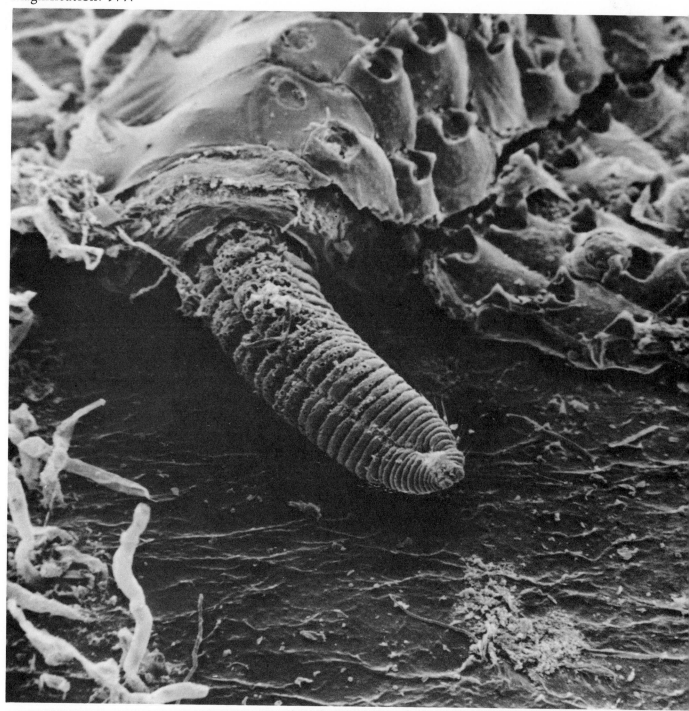

—and in part of the same microcosm shown in this scanning electron micrograph.

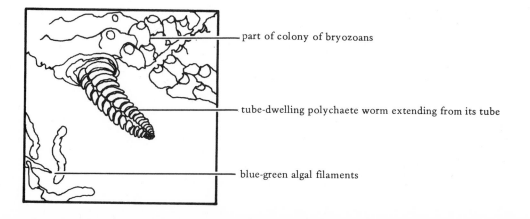

part of colony of bryozoans

tube-dwelling polychaete worm extending from its tube

blue-green algal filaments

This submerged portion from a fiberglass sailing vessel found as a derelict in the Atlantic Ocean has become a microbial garden.

Algal filaments form grass-like tufts on the left, while flower-like hydroids on stalks arise from stolons on the wood surface.

Algal filaments provide nutrients for their associated bacteria—

Plate 5-8. Microflora on wood trim
Site: North of Bermuda
Magnification: 92X

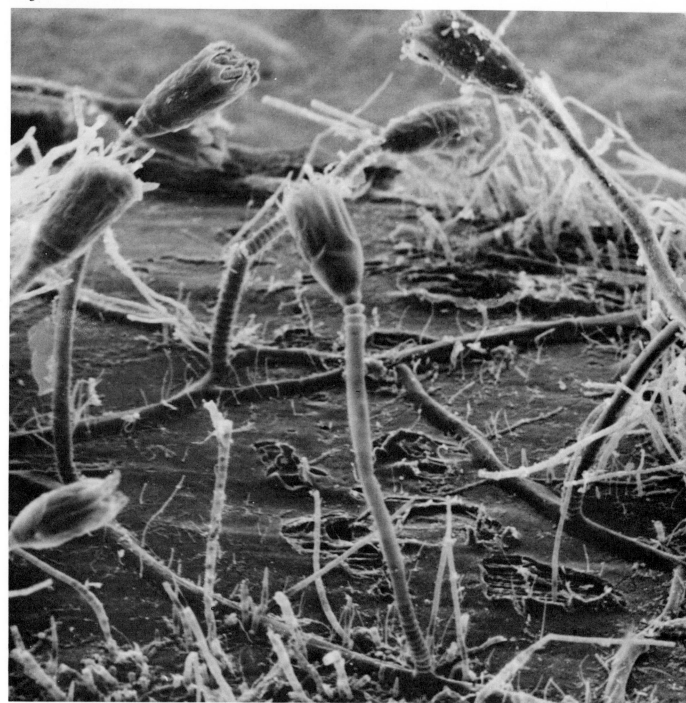

—while the campanularid hydroids use their tentacles to capture food particles.

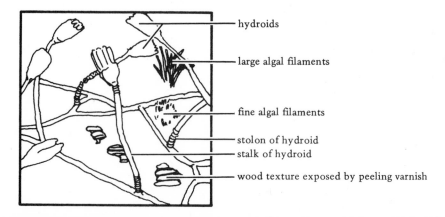

hydroids

large algal filaments

fine algal filaments

stolon of hydroid
stalk of hydroid

wood texture exposed by peeling varnish

*A submerged fiberglass surface from the same derelict sailboat provides an even better surface for settle-
ment, and therefore has a—*

—richer microcosm. Two goose barnacles as well as many hydroids make up the invertebrate fauna of this microcosm.

The fiberglass rods supply a textured surface that supports a garden of diatoms and bacteria—

Plate 5-11. Microflora on fiberglass fragment
Site: North of Bermuda
Magnification: 184X

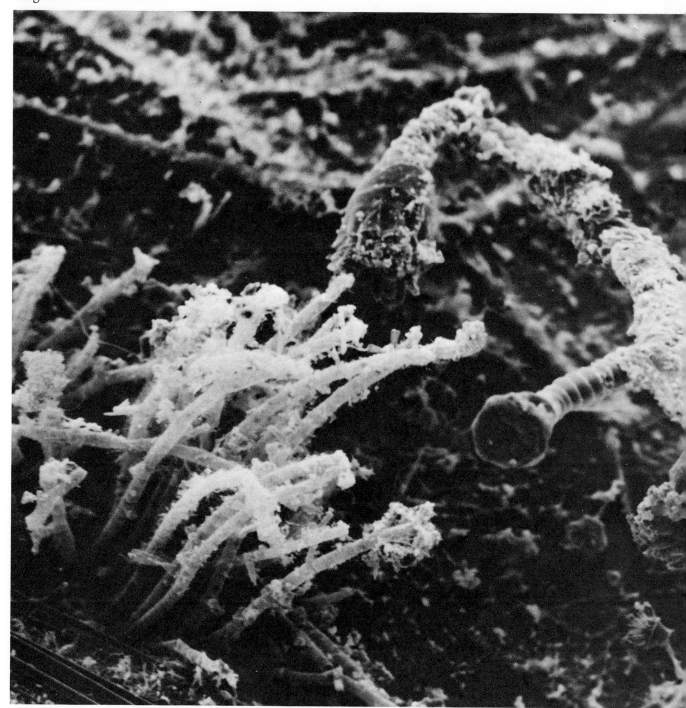

—from which arise filaments of blue-green algae and campanularid hydroids.

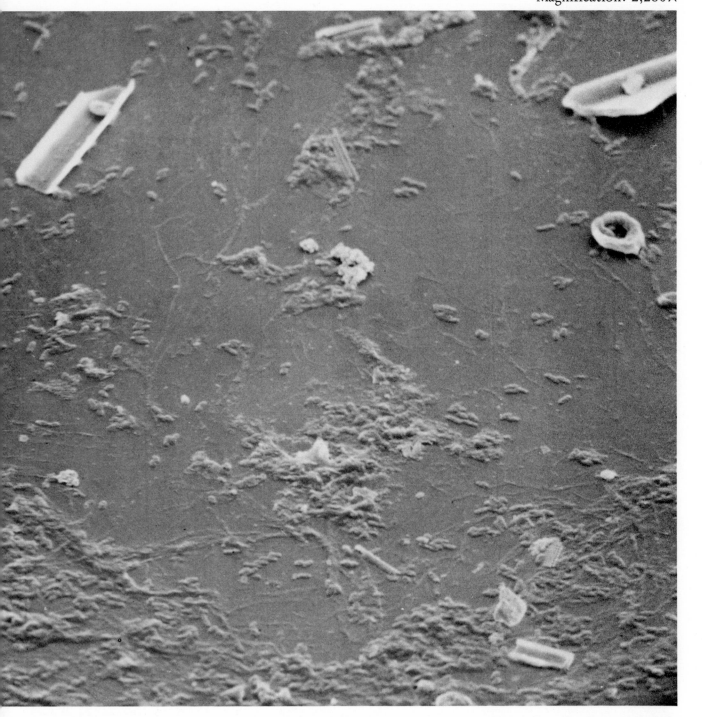

The smooth glass provides a poor surface for bacterial colonization and for examination by scanning electron microscopy—

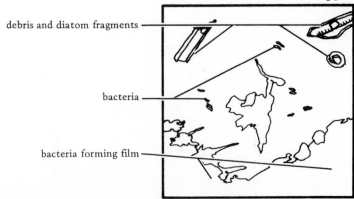

debris and diatom fragments

bacteria

bacteria forming film

Plate 5-13. Plastic cup immersed for four days
Site: Pigeon Key, FL
Magnification: 2,160X

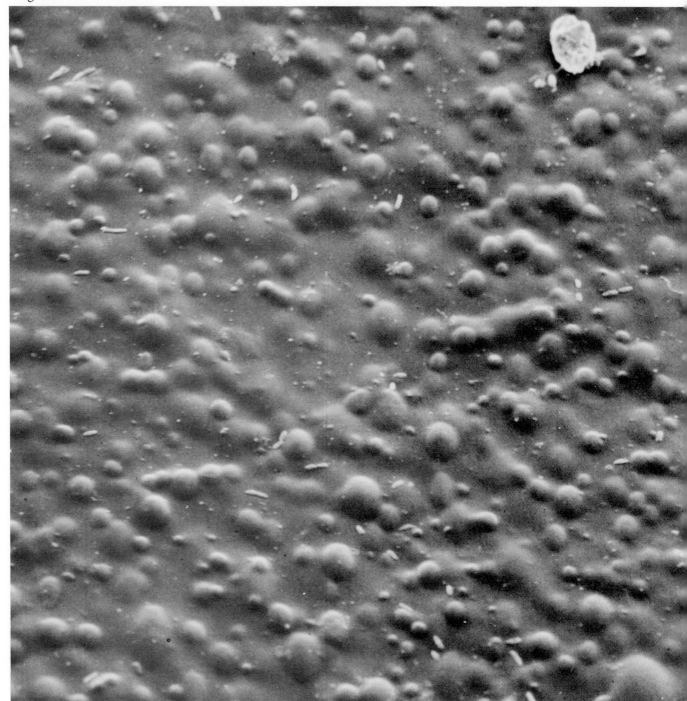

—while a textured plastic surface provides a better surface for colonization and examination.

Braided nylon provides a surface still better for the settlement of pennate diatoms—

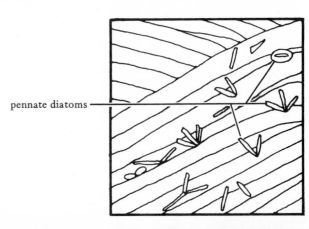

pennate diatoms

Plate 5-15. Nylon fishing line immersed for four days
Site: Pigeon Key, FL
Magnification: 2,975X

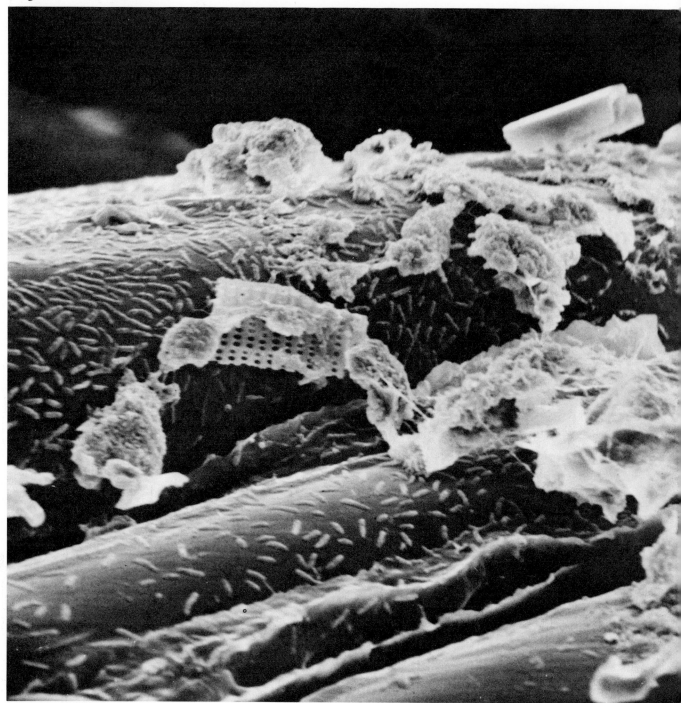

—*and the development of bacteria.*

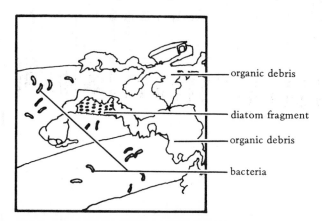

organic debris

diatom fragment

organic debris

bacteria

part six

ANIMAL SURFACES

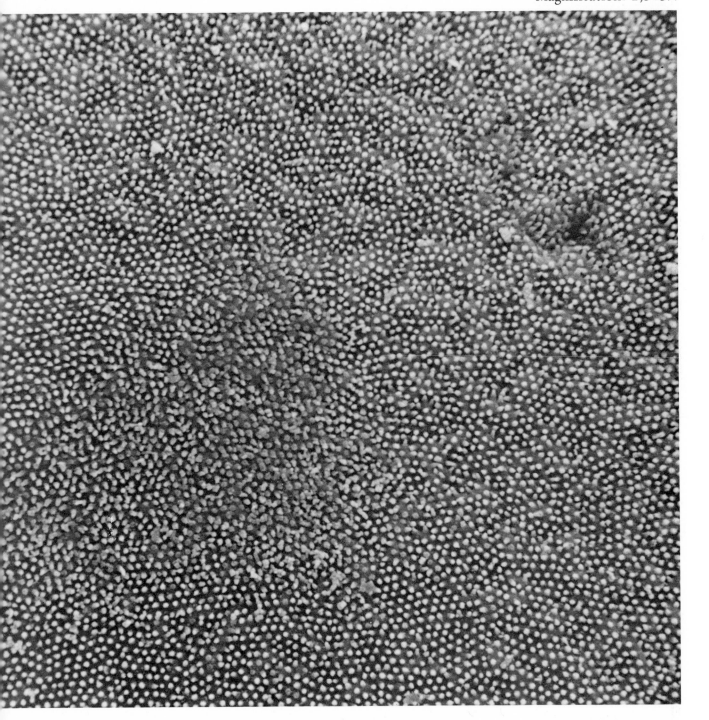

The freshly laid two-millimeter egg of the little sculpin Myoxocephalus aeneus *has a patterned texture relatively free of microorganisms and debris.*

animal surfaces

An assortment of animals that came to hand by chance encounter were photographed, collected, and examined. Because of their size and ease of capture and study, only the smaller invertebrates were used. Animal surfaces, like plant surfaces, can also enhance or inhibit microbial colonization. From the egg stage on, animals provide a surface for microbial attachment and colonization. Before the larvae hatch, a number of fish and crustacean eggs can be colonized. The eggs of a whelk take on a greenish-black hue which is caused by patches of algal filaments and diatoms and is visible at low magnifications. When the larvae hatch, a new surface is exposed. However, within 6 hours lobster larvae in culture already have bacteria attached to their newly exposed surface. By the 3rd day colonies of filamentous bacteria on the larvae are sending out their motile cells to colonize other surfaces. The colonization of new larval surfaces has not been studied in natural populations.

This potential for rapid and extensive colonization of animal surfaces often has been ignored in physiological experiments. Some investigators believe that certain invertebrates can utilize dissolved organic carbon as a source of nutrients (Anderson and Stephens, 1969). However, the well developed bacterial flora on many animals may not be completely controlled by antibiotic supplementation, and data on this phenomenon with non-germfree animals would be suspect. Under certain conditions, such as the intensive culture of larvae, algal and bacterial development can be so intense as to cause high mortalities, possibly through clogging of the gills. Most cases of epibiotic colonization seem to be examples of microorganisms taking the opportunity to colonize an empty space. There is some specialization, however, where certain microorganisms are found in certain habitats with special environmental conditions. Particle-ingesting ciliates seem to attach onto animals where there is an increased flow of water with its suspended particles. Unattached but crawling and grazing holotrichous ciliates seem to occur on highly nutrient surfaces which support a lawn of bacteria, such as the surface of the star coral *Astrangia danae*. Other surfaces support loricate and amoeboid protozoans.

In contrast to this microbial colonization that occurs on most animals, there is the remarkable absence of microorganisms on some others. The intact skin of fish, with its beautiful patterns of proteinaceous mucoid material (Plate 7-21), is one example. Other examples of freedom from an epibiotic microflora are a tunicate, a ship worm, and a nereid worm. The contrast between a good nutrient surface and one permitting only a slight selective colonization is shown by the hydroid *Obelia* growing on the surface of the brown alga *Laminaria* (Plates 6-70 to 6-73). An assortment of bacteria and protozoa live in a protected zone where the animal's root-like stolons branch across the plant surface. The animal stolons themselves are only sparsely colonized by distinctive forms. Such animal surfaces and their inhibitory or cleansing properties would bear further study.

The study of the associations of microorganisms with invertebrate animals has been limited mainly to the ciliates. Unfortunately, many of these associations have been considered to be parasitic rather than the commensal association they seem to be. The presence or absence of microorganisms epibiotic on animals may play an important role in the life of the animal. It seems logical that an adequate description of the epibiotic microbiota, at least that on the smaller forms, should become an integral part of the description and study of marine animals.

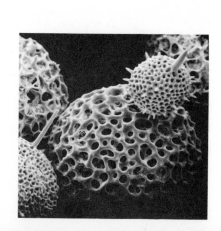

The discolorations on the tops of these incubating whelk eggs—

Plate 6-3. Egg of the dog whelk *Nucella (Thais) lapillis*
Site: Narragansett Bay, RI
Magnification: 53X

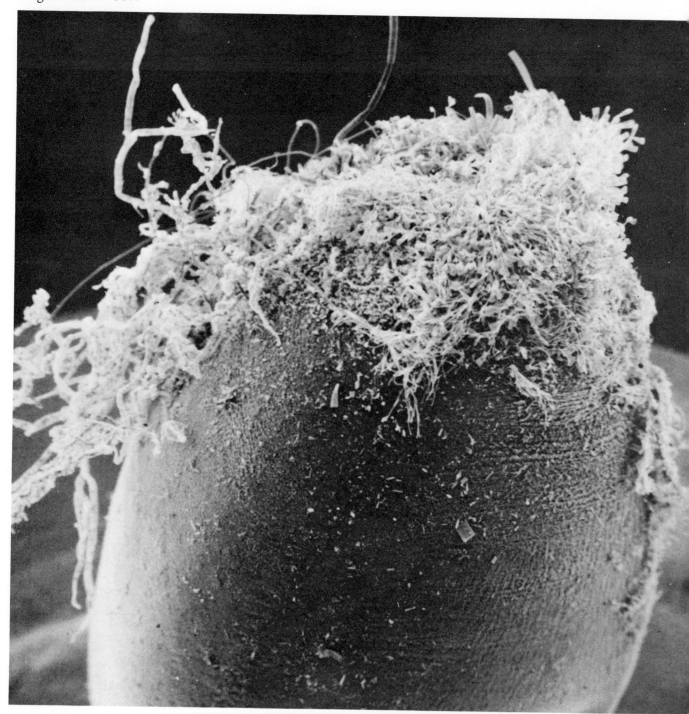

—are algal accumulations—

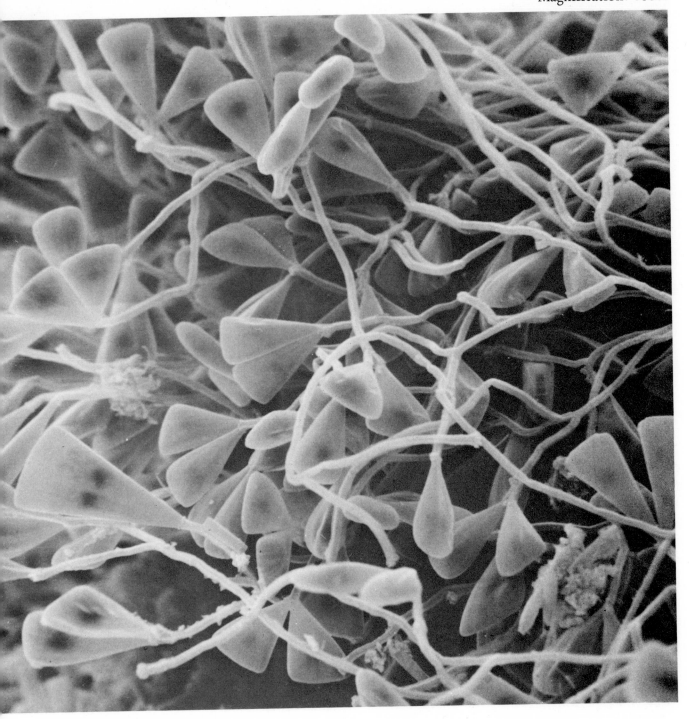

—composed of the stalked diatom Licmophora*—*

Plate 6-5. Surface of whelk egg
Site: Narragansett Bay, RI
Magnification: 5,230X

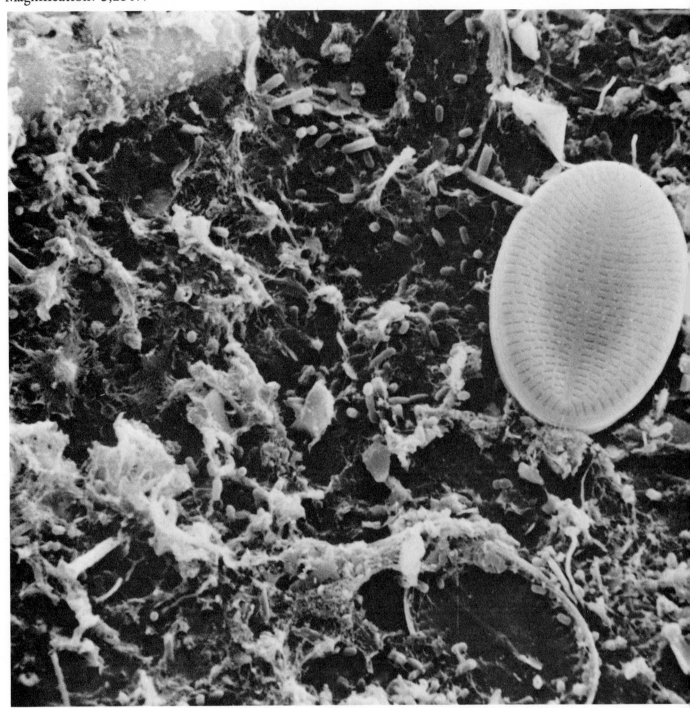

—*while the egg surface below them is encrusted with debris, bacteria, and pennate diatoms.*

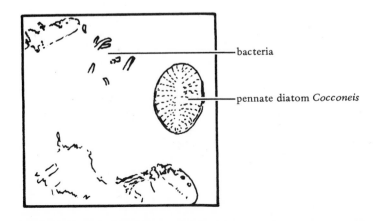

bacteria

pennate diatom *Cocconeis*

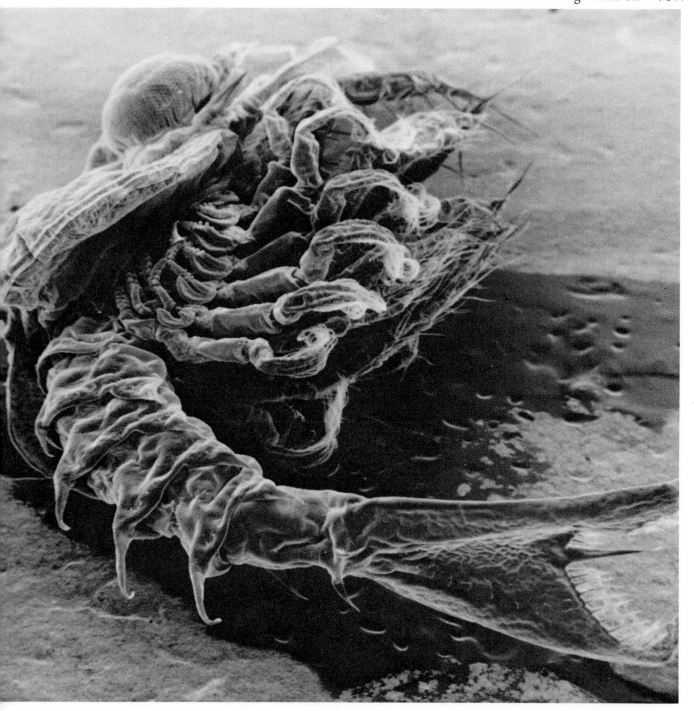

The newly hatched and wrinkled lobster immediately becomes colonized with bacteria—

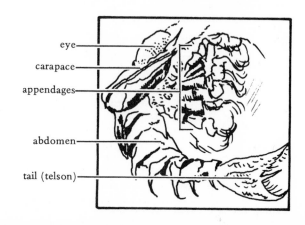

Plate 6-7. Carapace of three-day-old lobster
Site: University of Rhode Island Aquaculture Project
Magnification: 2,600×

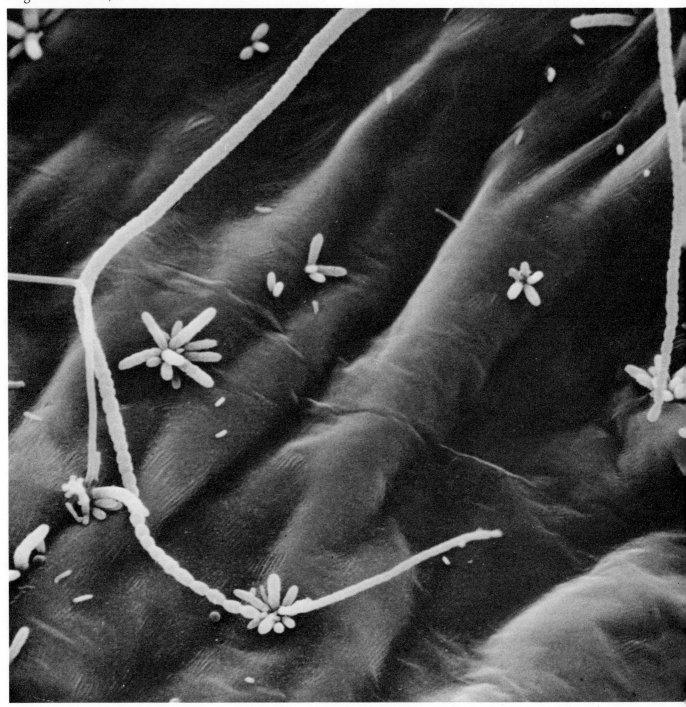

—and within days shows the different stages in the life cycle of the filamentous bacterium Leucothrix mucor—

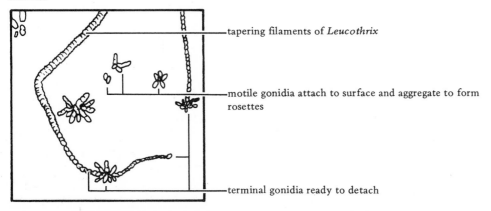

tapering filaments of *Leucothrix*

motile gonidia attach to surface and aggregate to form rosettes

terminal gonidia ready to detach

—which form tufts of filaments—

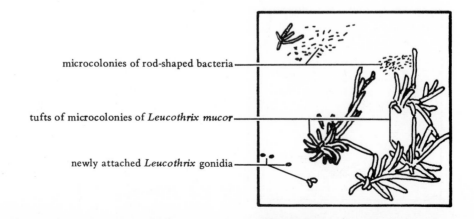

microcolonies of rod-shaped bacteria

tufts of microcolonies of *Leucothrix mucor*

newly attached *Leucothrix* gonidia

—*that extend from the animal surface.*

These distinctive caprellid crustaceans—

Plate 6-11. Amphipod *Caprella grahamii*
Site: Narragansett Bay, RI
Magnification: 54X

—can be heavily colonized—

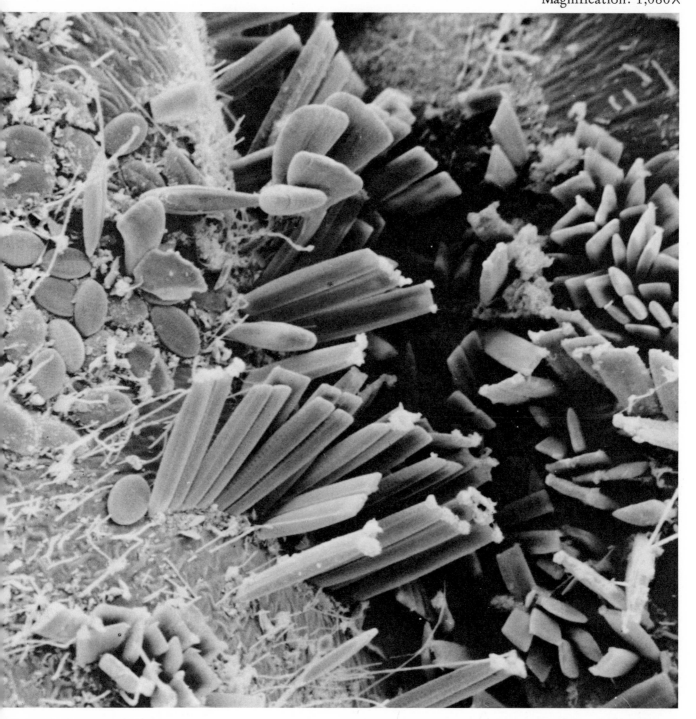

—by a mixture of pennate diatoms and bacterial filaments.

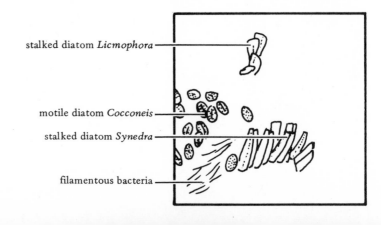

stalked diatom *Licmophora*

motile diatom *Cocconeis*

stalked diatom *Synedra*

filamentous bacteria

Plate 6-13. Tube-dwelling amphipod *Jassa falcata*
Site: Narragansett Bay, RI
Magnification: 10X

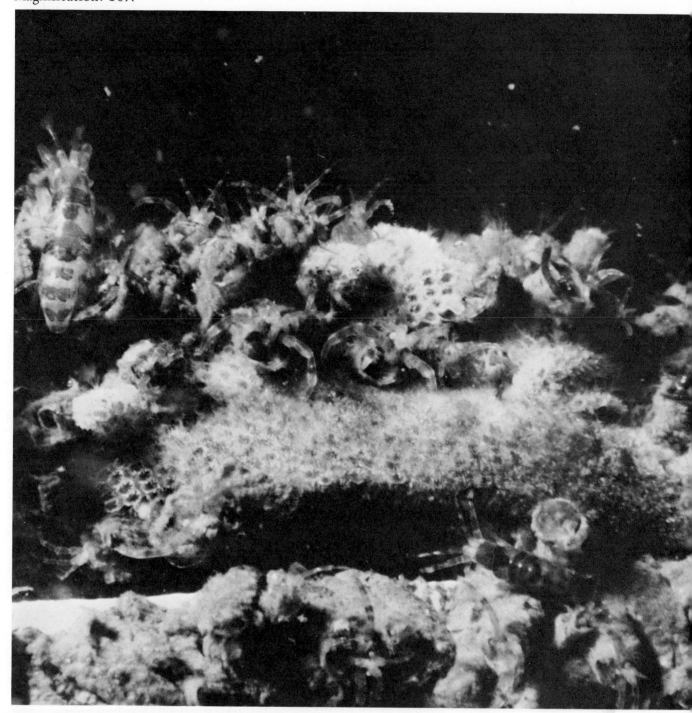

These amphipods, approximately five millimeters long, live in tubes constructed of debris.

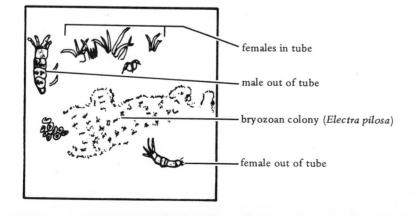

— females in tube

— male out of tube

— bryozoan colony (*Electra pilosa*)

— female out of tube

The amphipod tubes attached to the holdfast of a seaweed (Laminaria)—

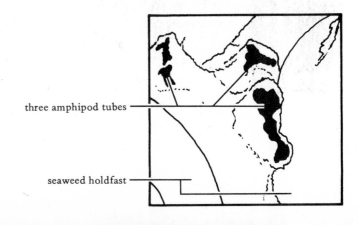

three amphipod tubes

seaweed holdfast

Plate 6-15. Microbiota of amphipod tube
Site: Narragansett Bay, RI
Magnification: 1,520X

—*are heavily colonized by filamentous bacteria and a few pennate diatoms (*Grammatophora*).*

The lighter areas on the exoskeleton of the amphipod—

Plate 6-17. Carapace of the amphipod *Jassa falcata*
Site: Narragansett Bay, RI
Magnification: 2,830×

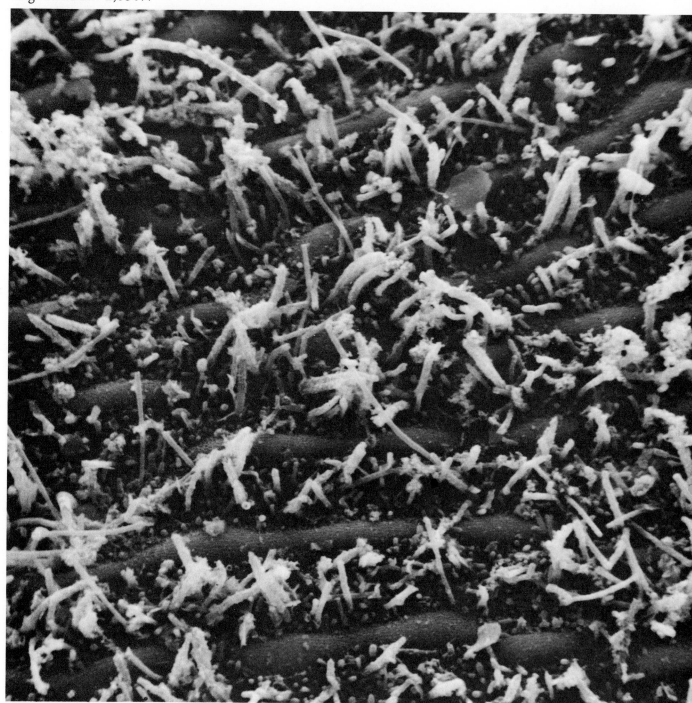

—are also caused by bacteria.

A lateral view of this odd creature walking over a colony of campanularid hydroids.

Plate 6-19. Head and claws of sea spider
Site: Narragansett Bay, RI
Magnification: 195X

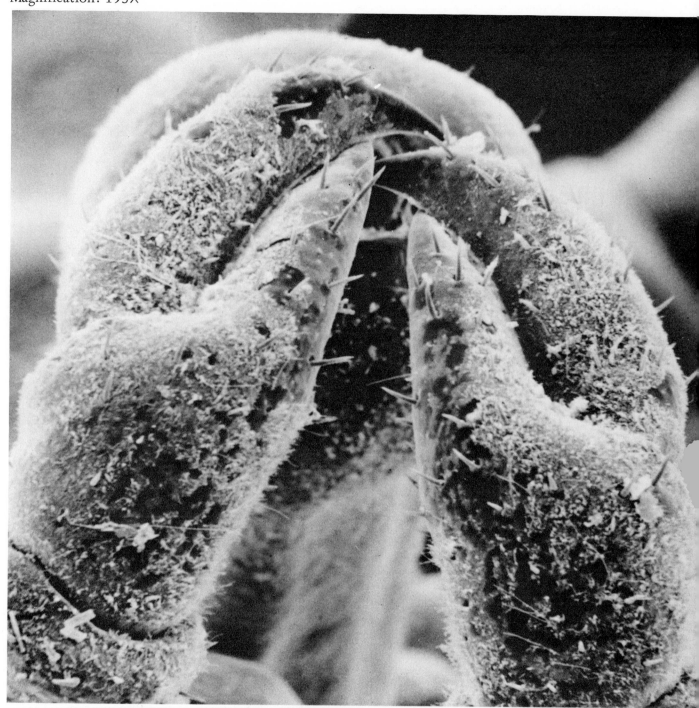

The claws are heavily fouled—

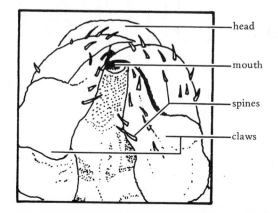

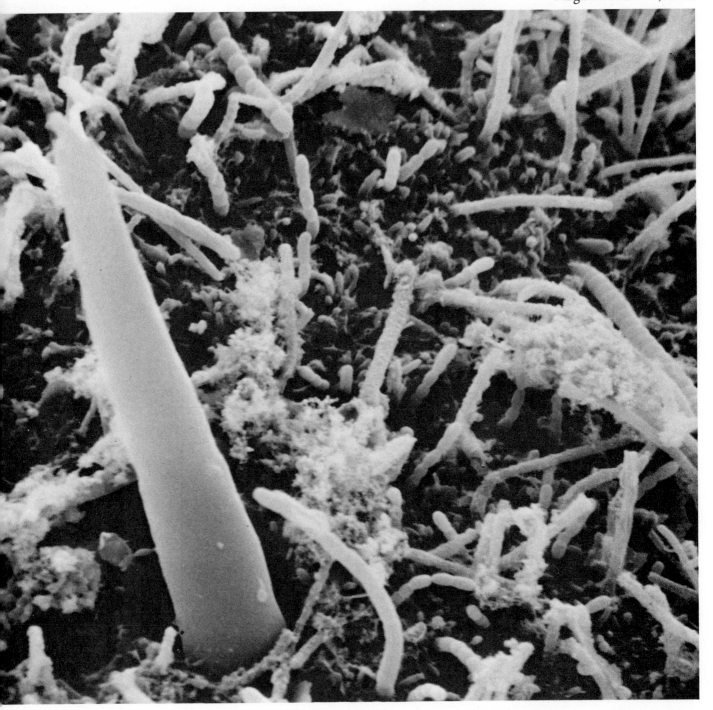

—by the filamentous bacterium Leucothrix mucor *and other bacteria—*

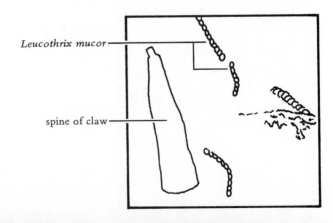

Leucothrix mucor

spine of claw

Plate 6-21. Dorsal surface of sea spider
Site: Narragansett Bay, RI
Magnification: 3,355X

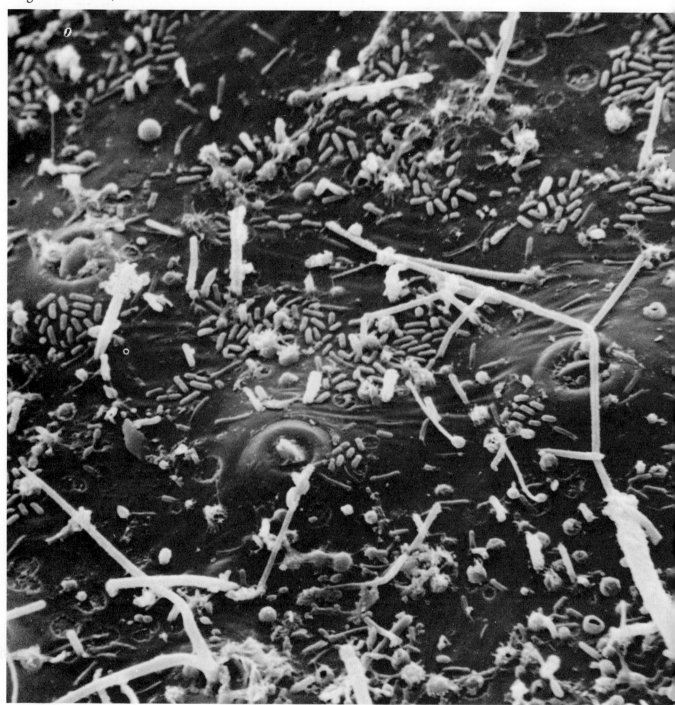

—while the back of the body is more lightly colonized by microcolonies of rod-shaped bacteria and a few bacterial filaments.

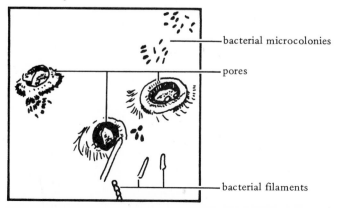

bacterial microcolonies

pores

bacterial filaments

Dorsal view of a scaleworm, ten millimeters long, crawling over rock surface.

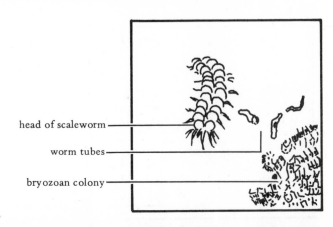

head of scaleworm ——

worm tubes ——

bryozoan colony ——

Plate 6-23. Scaleworm *Harmothoe imbricata*
Site: Narragansett Bay, RI
Magnification: 50X

Frontal view of same scaleworm.

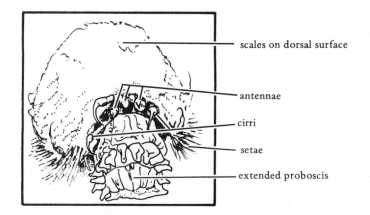

scales on dorsal surface

antennae

cirri

setae

extended proboscis

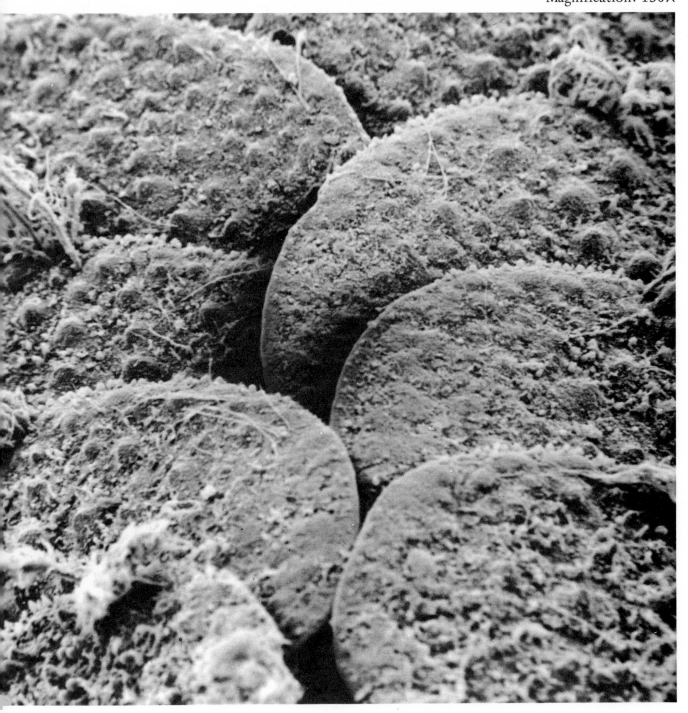

The textured surface of the scales is heavily encrusted—

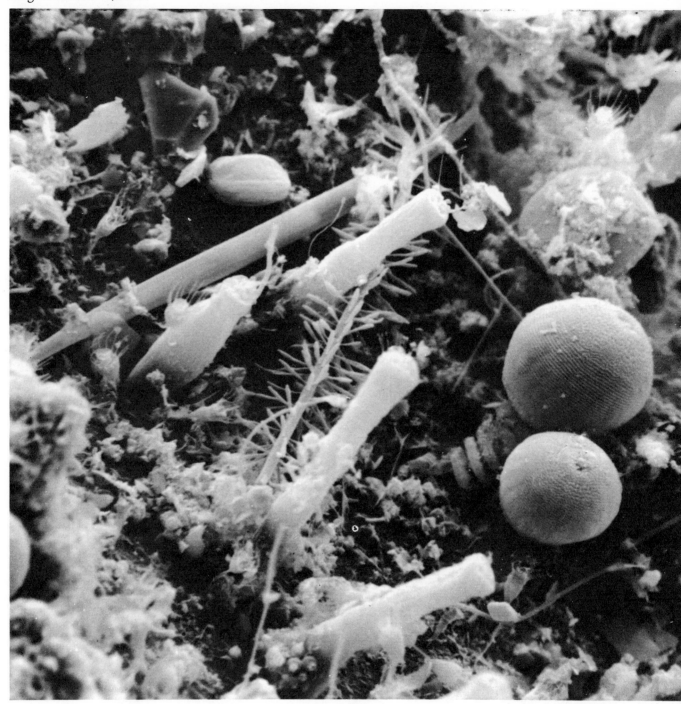

—by a mixture of microbial forms—

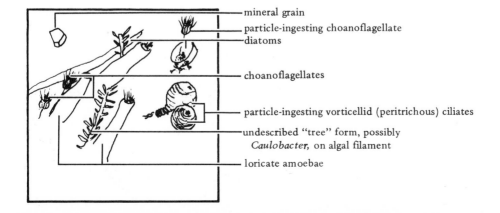

mineral grain
particle-ingesting choanoflagellate
diatoms
choanoflagellates
particle-ingesting vorticellid (peritrichous) ciliates
undescribed "tree" form, possibly
   Caulobacter, on algal filament
loricate amoebae

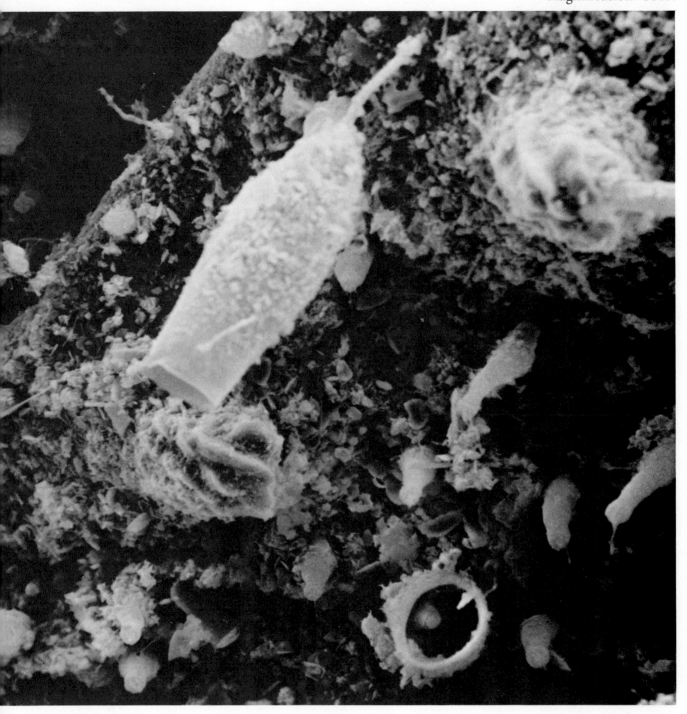

—including ciliated protozoans—

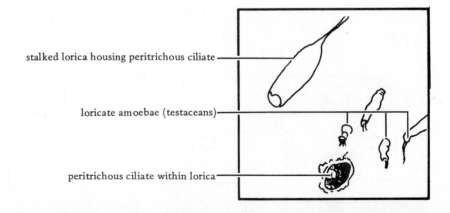

stalked lorica housing peritrichous ciliate

loricate amoebae (testaceans)

peritrichous ciliate within lorica

Plate 6-27. Loricate peritrichous ciliate on scaleworm
Site: Narragansett Bay, RI
Magnification: 3,700×

—that live within a protective lorica or house.

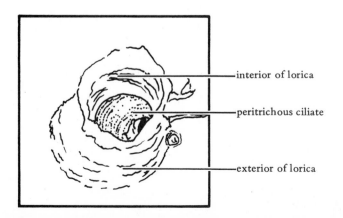

interior of lorica

peritrichous ciliate

exterior of lorica

The rough surface of a limpet's shell—

Plate 6-29. Microflora on limpet shell
Site: Narragansett Bay, RI
Magnification: 2,700×

—is littered with debris, diatoms, and bacterial filaments—

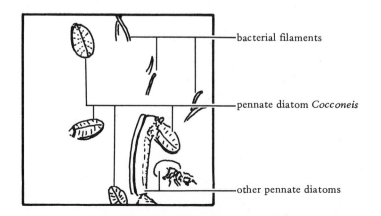

bacterial filaments

pennate diatom *Cocconeis*

other pennate diatoms

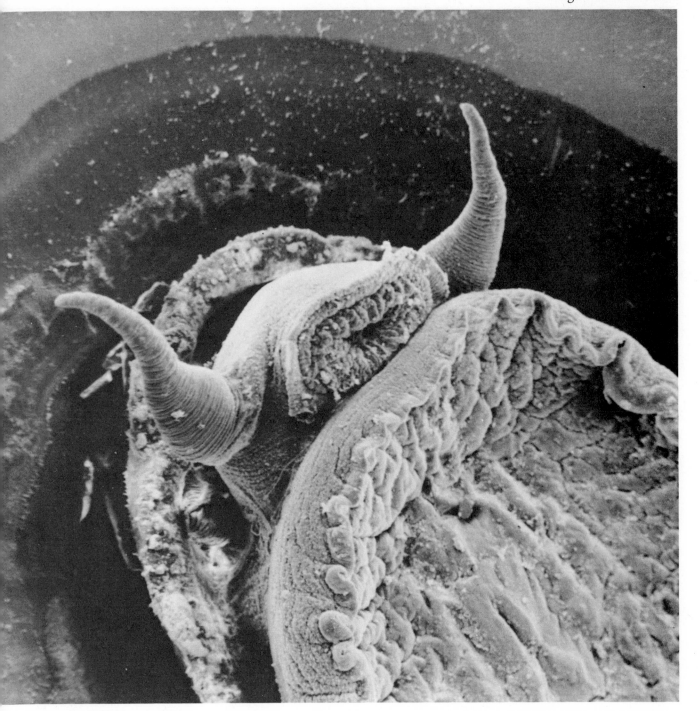

—while the animal itself is quite clean—

—except for these different sized spheres (associated with the host's cilia) that have the external appearance of thecate colonies of ammonia-oxidizing bacteria.

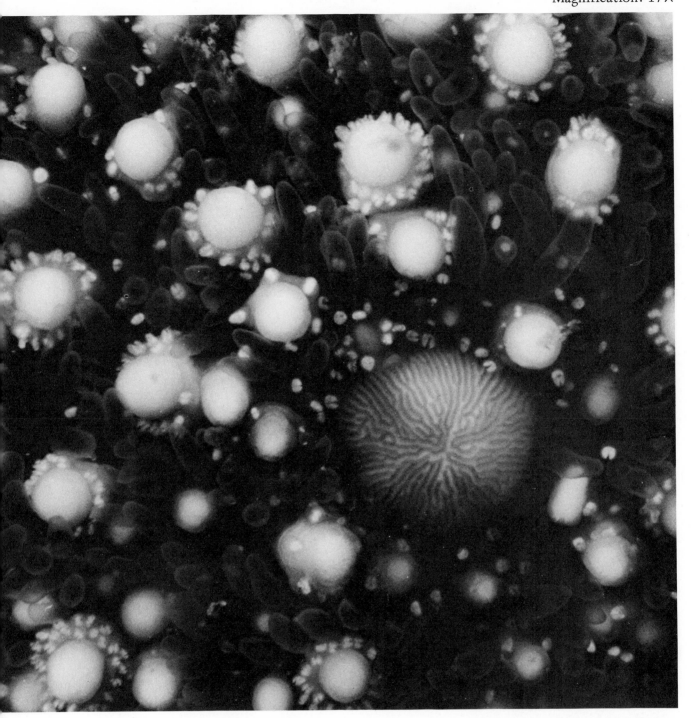

The area around the four-millimeter water intake pore of this twenty-five-centimeter starfish—

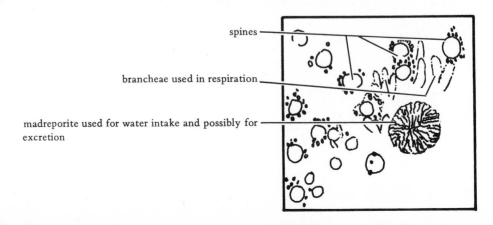

spines

brancheae used in respiration

madreporite used for water intake and possibly for excretion

Plate 6-33. Madreporite area of starfish *Asterias forbesi*
Site: Narragansett Bay, RI
Magnification: 60X

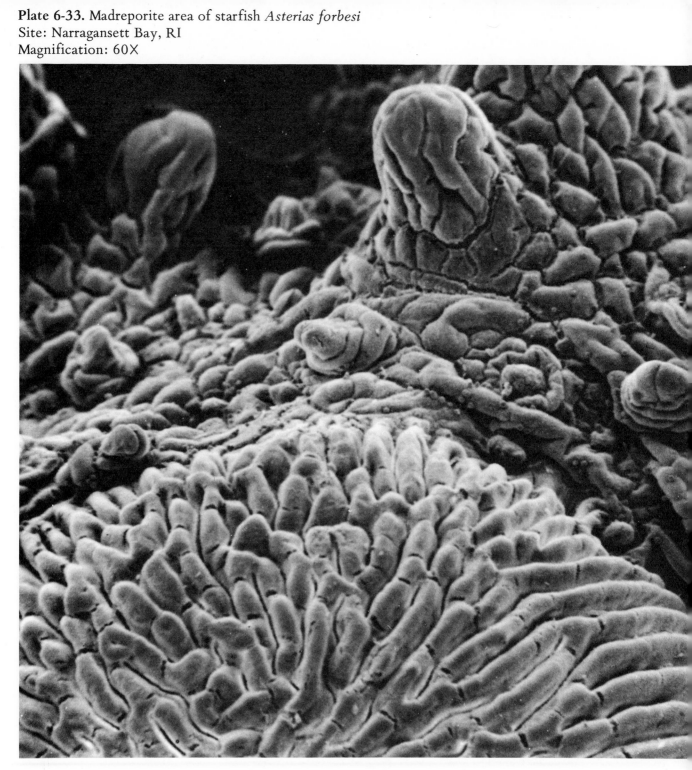

—loses its translucent beauty in scanning electron microscopy—

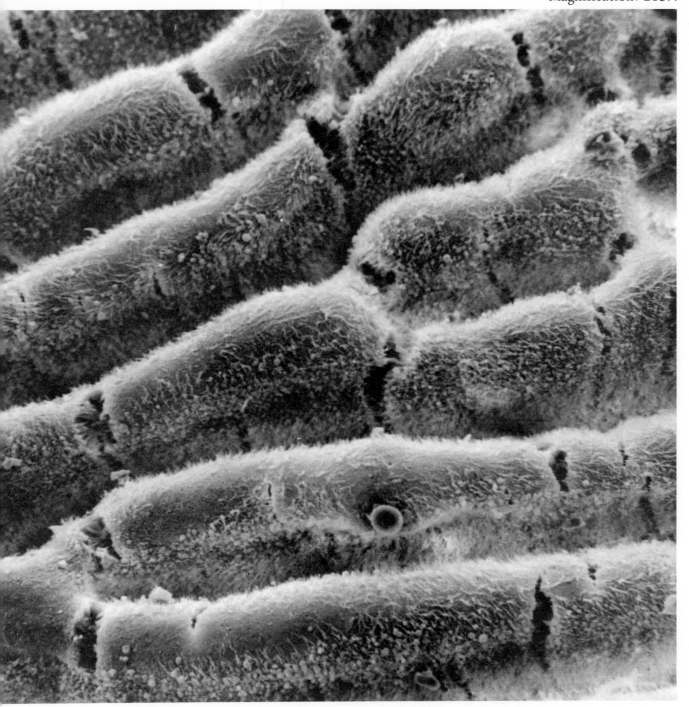

—but reveals the ciliated nature of the pore—

Plate 6-35. Microbiota of starfish madreporite
Site: Narragansett Bay, RI
Magnification: 1,400X

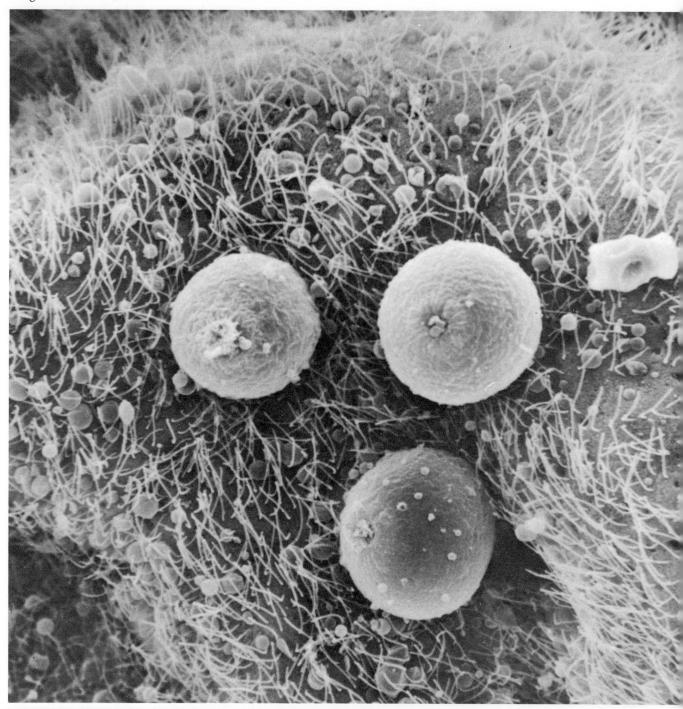

—that contains vorticellid protozoans and spheres suggestive of thecate colonies of ammonia-oxidizing bacteria.

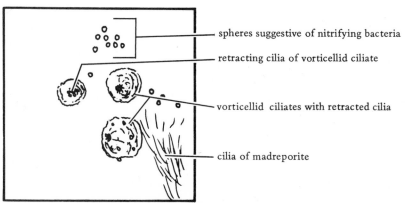

spheres suggestive of nitrifying bacteria

retracting cilia of vorticellid ciliate

vorticellid ciliates with retracted cilia

cilia of madreporite

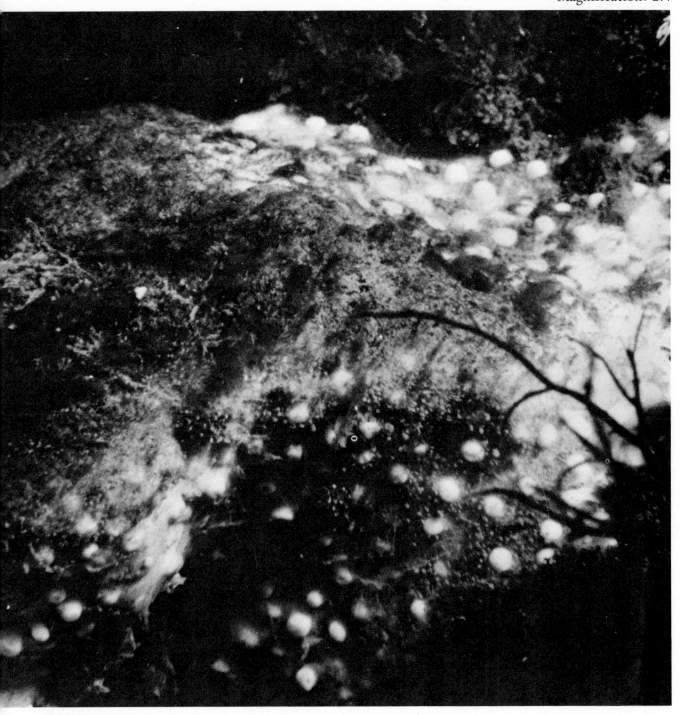

This boring sponge, which decays dead shells on the sea floor, has some of its yellow papillae exposed, while much of its surface—

Plate 6-37. Encrusted surface of sulphur sponge
Site: Narragansett Bay, RI
Magnification: 1,000X

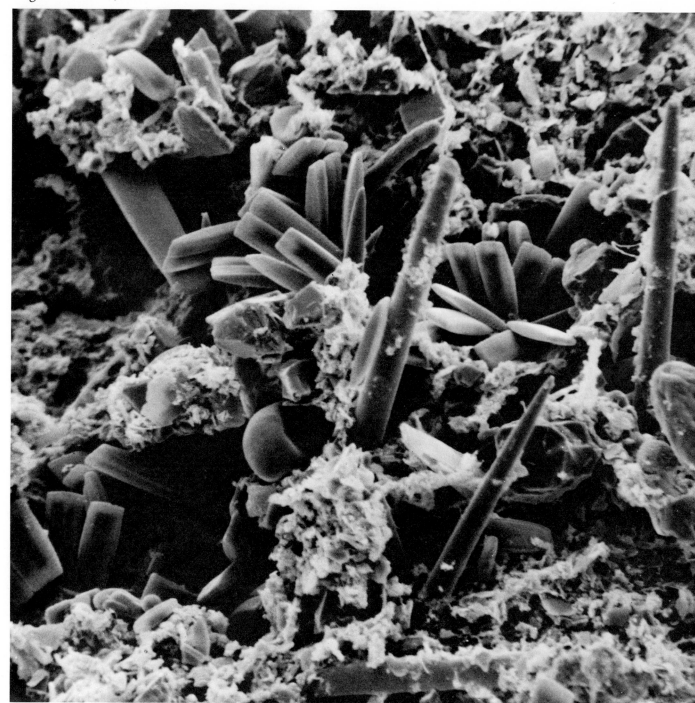

—is encrusted with pennate diatoms and debris which obscure the surface and all but cover the spicules—

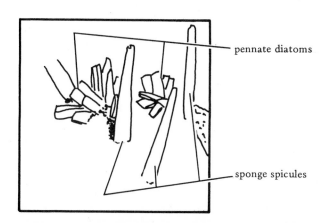

pennate diatoms

sponge spicules

—while the exposed spicule surface of its warm water cousin—

Plate 6-39. Microbiota of red sponge
Site: Molasses Key, FL
Magnification: 5,000X

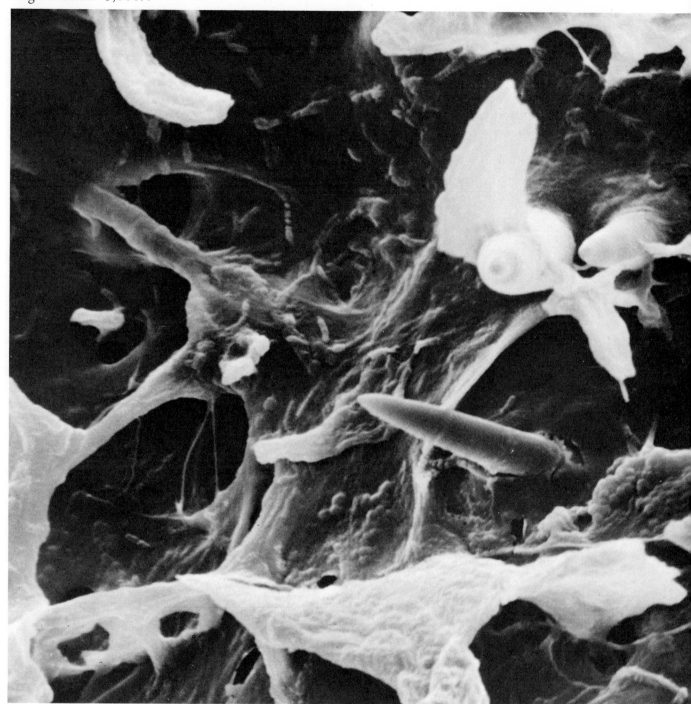

—contains a bacteria-rich slime characteristic of these porous animals.

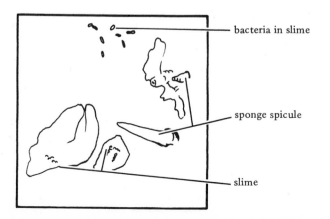

bacteria in slime

sponge spicule

slime

The translucent polyp with its extended tentacles—

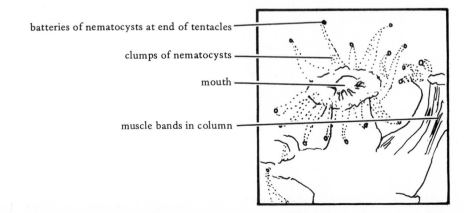

batteries of nematocysts at end of tentacles

clumps of nematocysts

mouth

muscle bands in column

Plate 6-41. Oral view of polyp of the northern star coral *Astrangia danae*
Site: Narragansett Bay, RI
Magnification: 52X

—is colonized, even when retracted, by a herd of hovering particles just visible here—

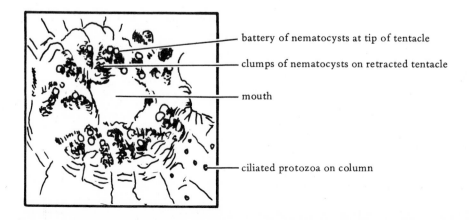

battery of nematocysts at tip of tentacle

clumps of nematocysts on retracted tentacle

mouth

ciliated protozoa on column

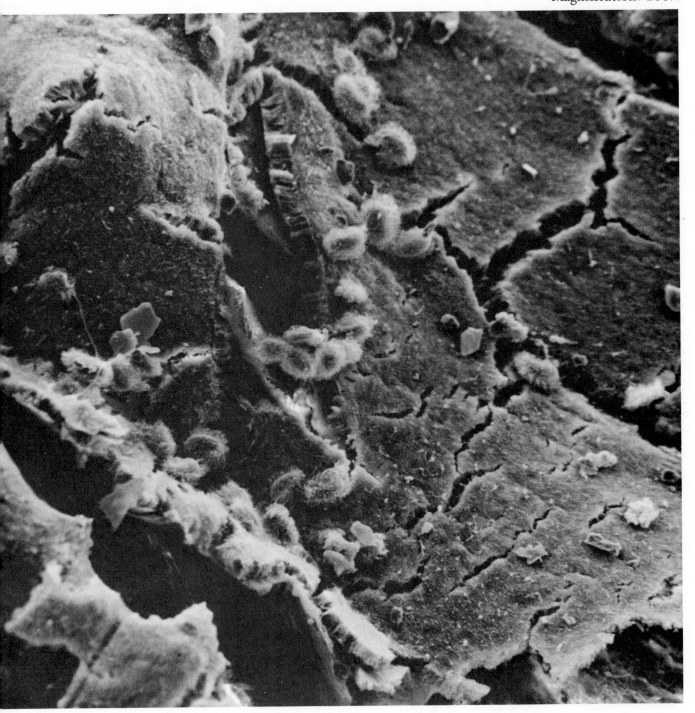

—which on closer examination are recognizable as motile, ciliated protozoans—

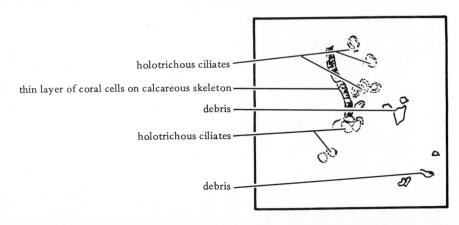

holotrichous ciliates

thin layer of coral cells on calcareous skeleton

debris

holotrichous ciliates

debris

Plate 6-43. Surface of coral polyp column
Site: Narragansett Bay, RI
Magnification: 1,425X

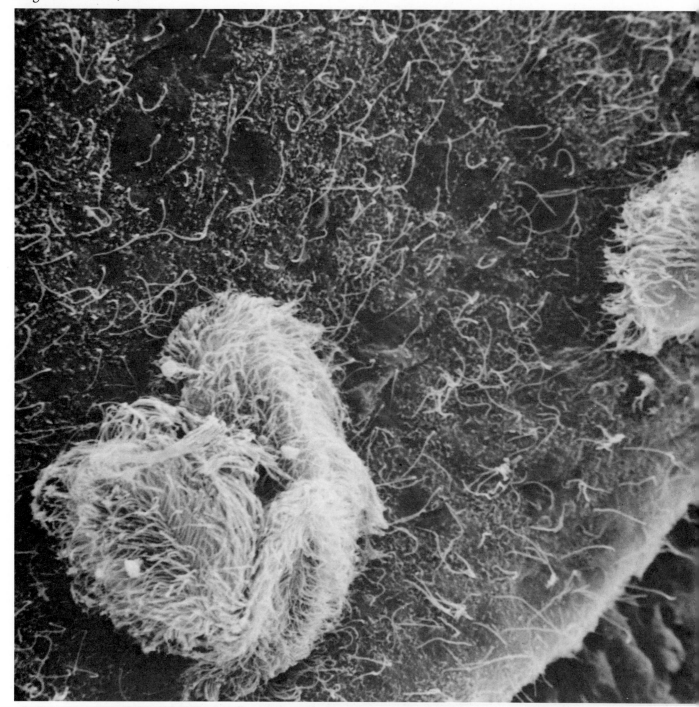

—*that seem to be grazing on a lawn of bacteria on the coral surface and leaving bare areas.*

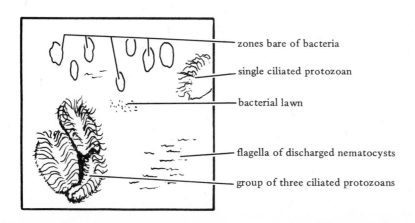

zones bare of bacteria

single ciliated protozoan

bacterial lawn

flagella of discharged nematocysts

group of three ciliated protozoans

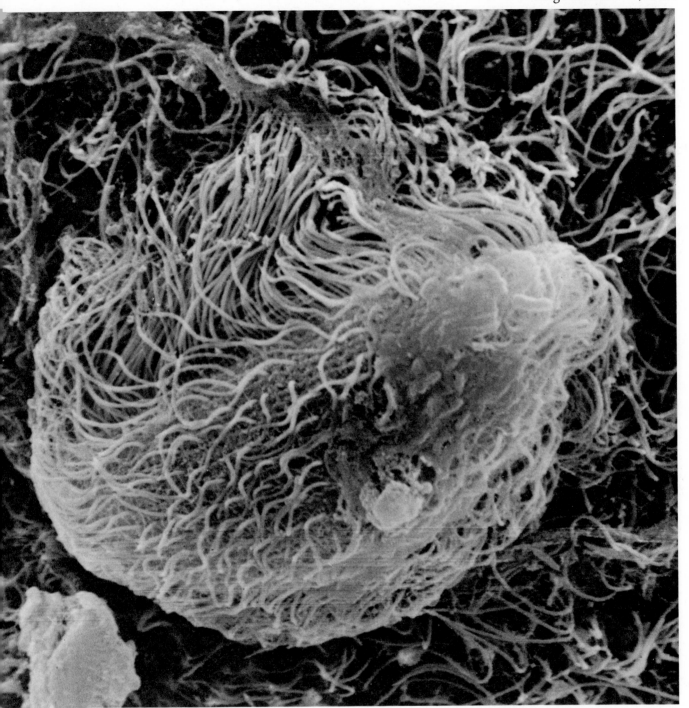

The hairy looking holotrichous ciliate—

—*when denuded by chloral hydrate, reveals the basal bodies (kinetosomes) of the cilia.*

The water currents set up by the feeding action of this flower-like animal—

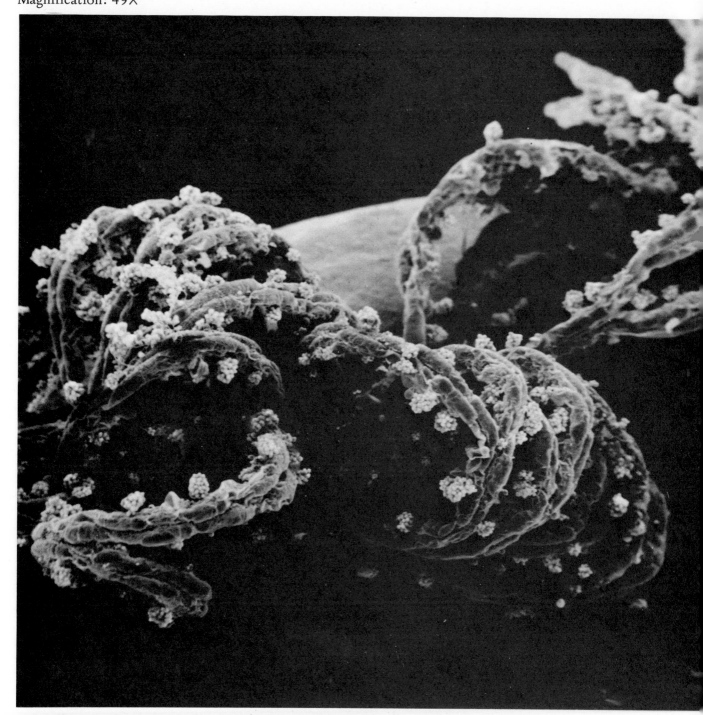

—provide an ideal environment for the development of grape-like clusters—

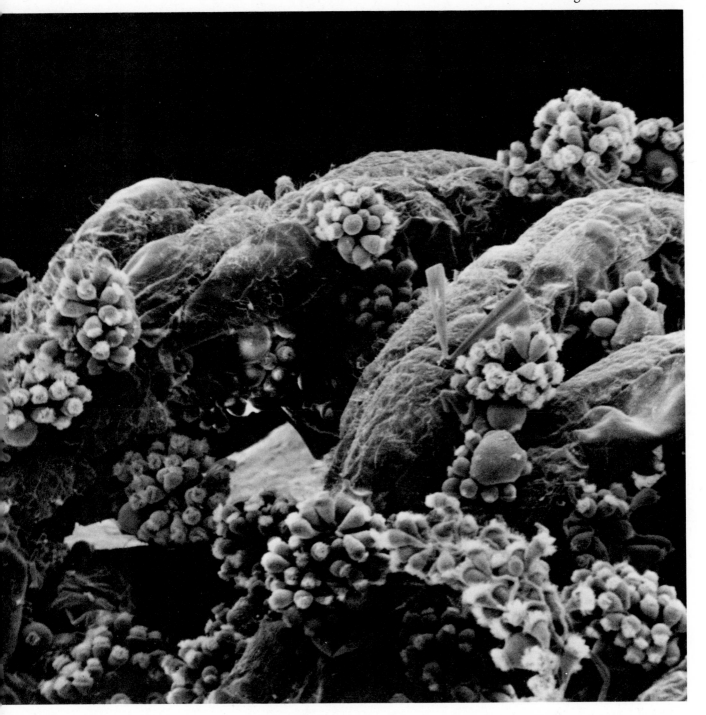

—of the colonial ciliate Zoothamnium—

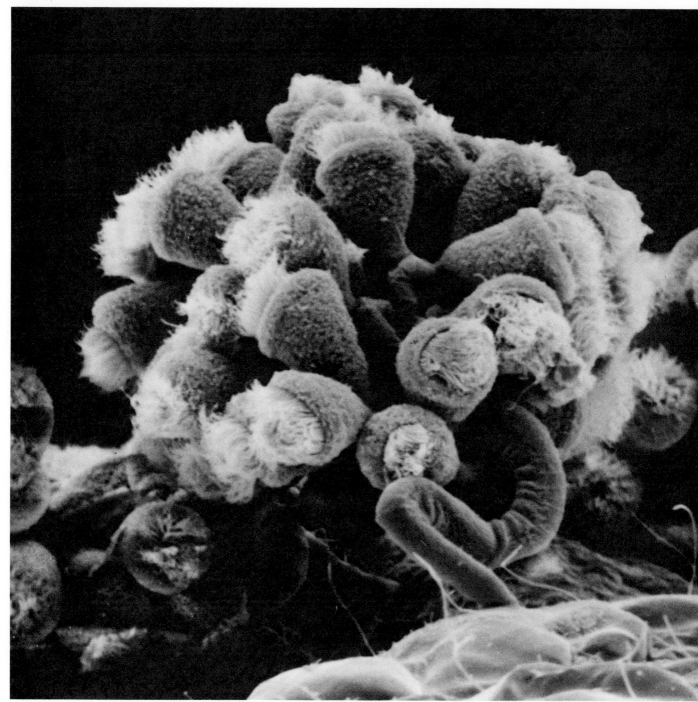

—which uses its tufts of cilia to ingest particles suspended in the water currents set up by its bryozoan host.

The textured surface of these protozoans—

Plate 6-51. Surface of *Zoothamnium*
Site: Narragansett Bay, RI
Magnification: 6,000X

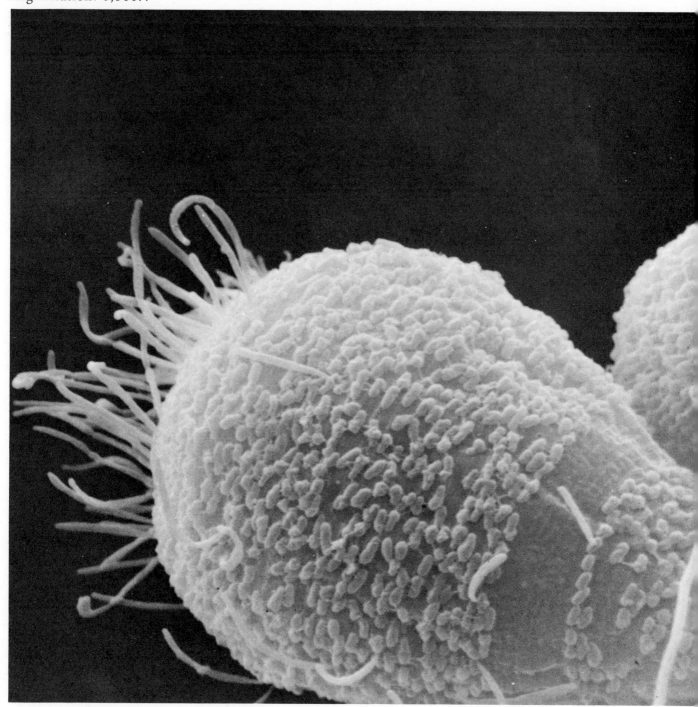

—is caused by a lawn of rod-shaped bacteria.

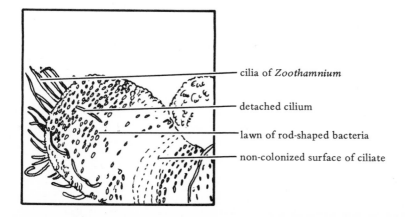

cilia of *Zoothamnium*

detached cilium

lawn of rod-shaped bacteria

non-colonized surface of ciliate

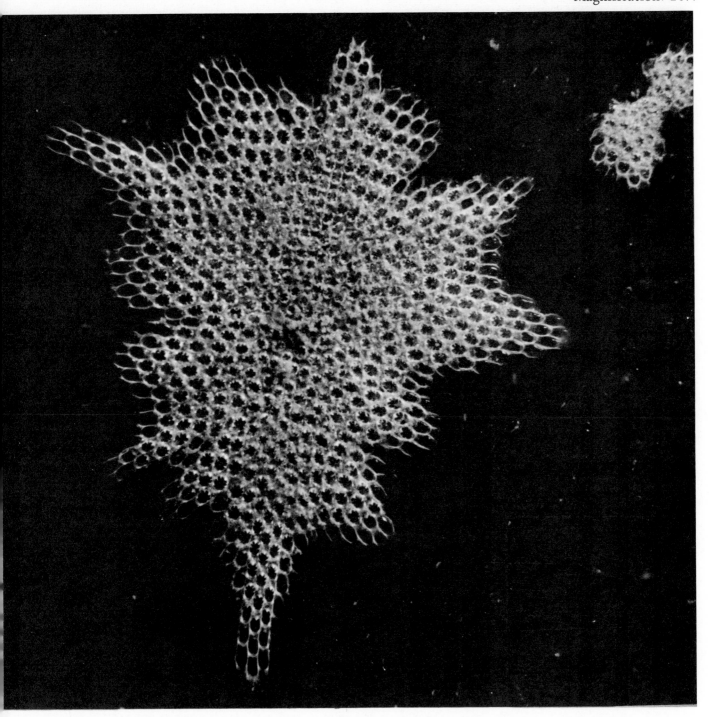

Other bryozoans form flat colonies on a variety of surfaces—

Plate 6-53. Edge of colony of bryozoan *Alcyonidium hirsutum*
Site: Narragansett Bay, RI
Magnification: 126X

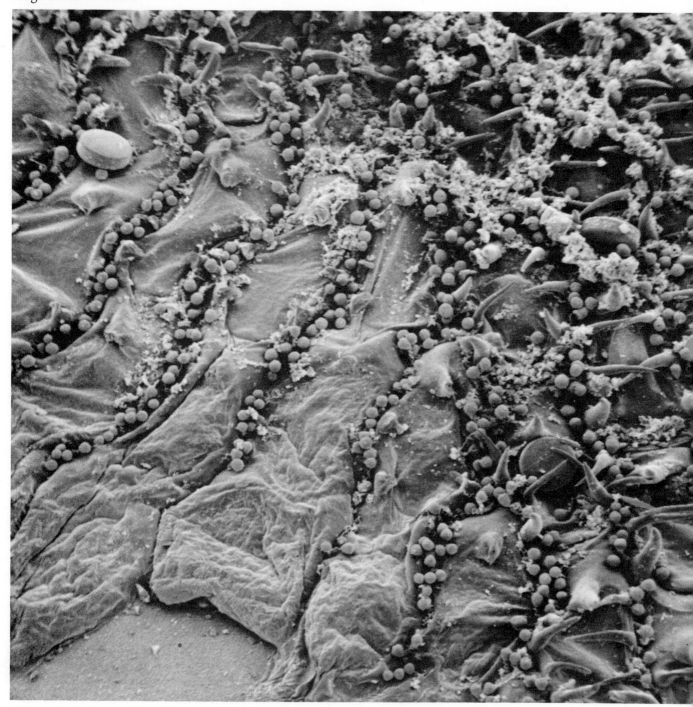

—*where the water currents set up by their feeding activities also support rich populations of attached protozoa*—

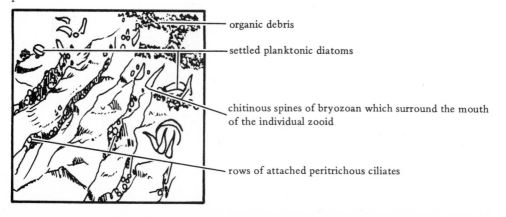

organic debris

settled planktonic diatoms

chitinous spines of bryozoan which surround the mouth of the individual zooid

rows of attached peritrichous ciliates

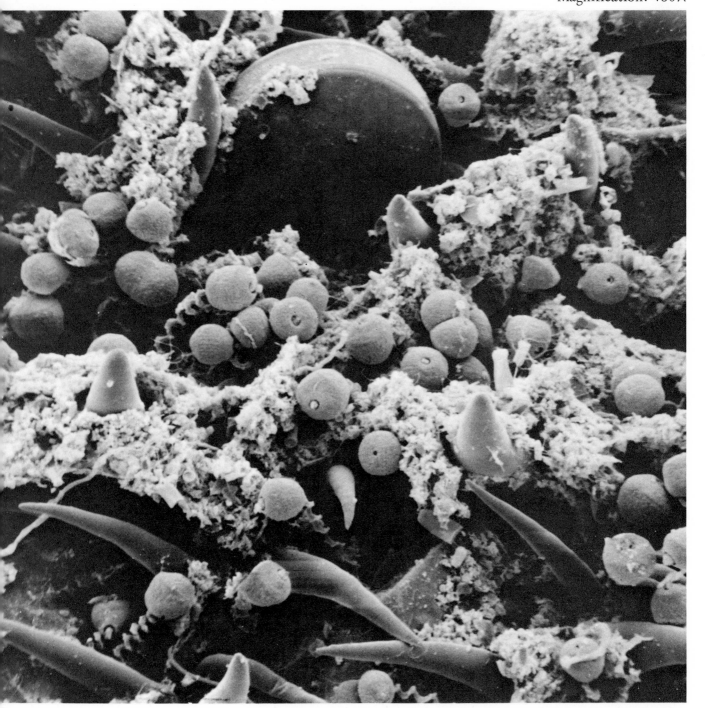

—consisting of the individual ciliated protozoan Vorticella *with its spiral stalk.*

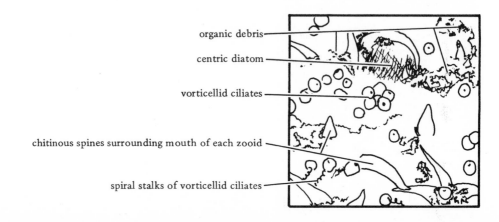

organic debris
centric diatom
vorticellid ciliates
chitinous spines surrounding mouth of each zooid
spiral stalks of vorticellid ciliates

Plate 6-55. Vorticellid ciliates on bryozoan *Alcyonidium hirsutum*
Site: Narragansett Bay, RI
Magnification: 1,335X

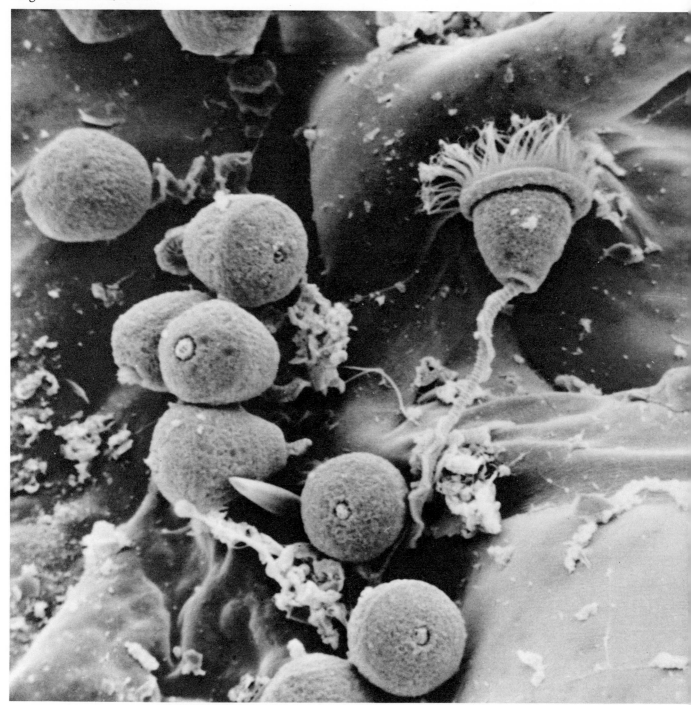

A ciliate with an extended contractile stalk and extended oral ciliature.

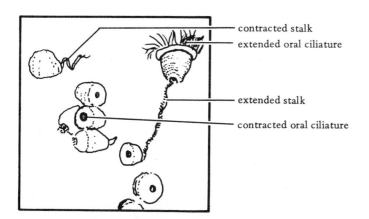

contracted stalk

extended oral ciliature

extended stalk

contracted oral ciliature

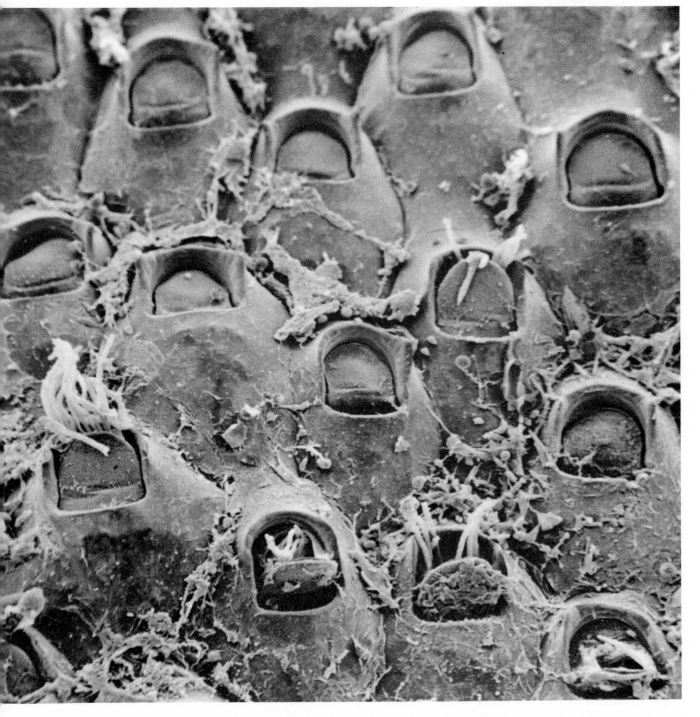

*This lightly colonized area shows several individual zooids within the colony with open "trap doors"
(opercula) and extended tentacles—*

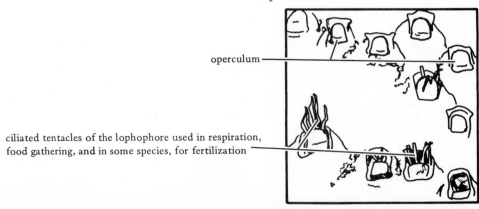

operculum —————

ciliated tentacles of the lophophore used in respiration,
food gathering, and in some species, for fertilization

Plate 6-57. Ascophoran bryozoan on eelgrass
Site: Narragansett Bay, RI
Magnification: 250X

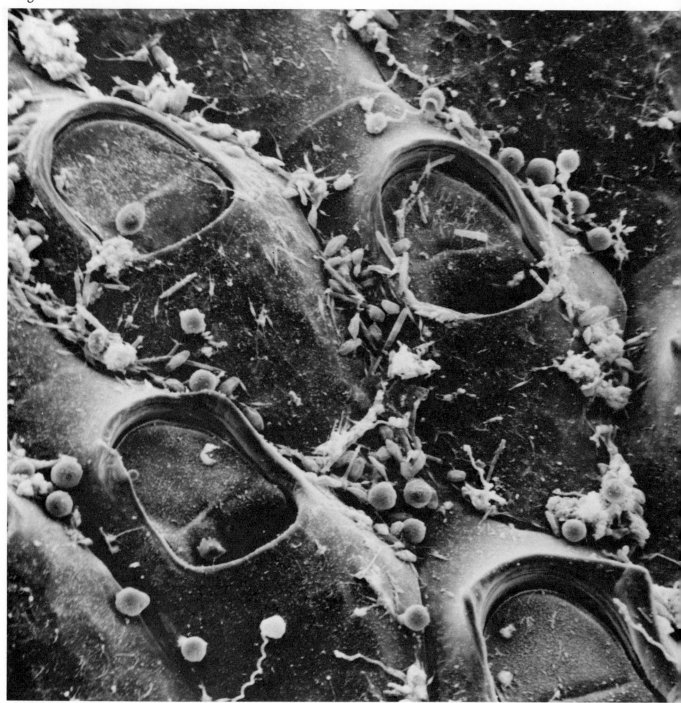

—*that attract and sustain the attached protozoans with their suspended particles.*

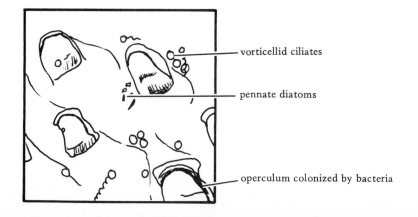

vorticellid ciliates

pennate diatoms

operculum colonized by bacteria

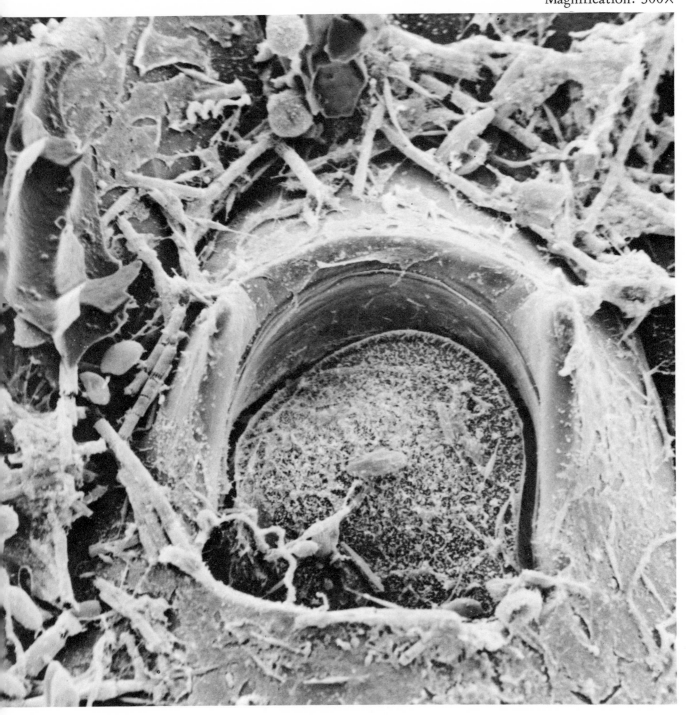

The bryozoan also provides a surface for attached pennate diatoms and settled planktonic diatoms—

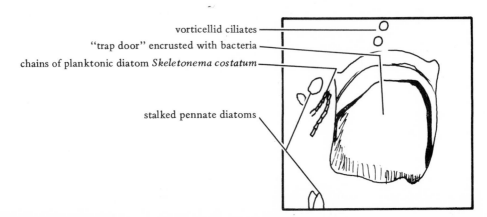

vorticellid ciliates ——

"trap door" encrusted with bacteria ——

chains of planktonic diatom *Skeletonema costatum* ——

stalked pennate diatoms ——

Plate 6-59. "Trap door" (operculum) of an ascophoran bryozoan
Site: Narragansett Bay, RI
Magnification: 1,580X

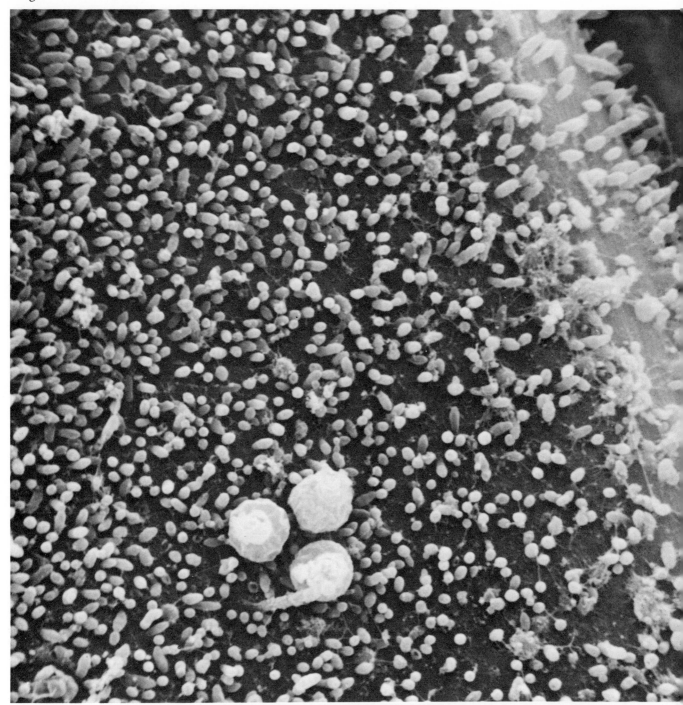

—as well as a rich bacterial lawn and loricate amoebae (testaceans)

Plate 6-60. Hydroid and bryozoan on pelagic gulfweed *Sargassum natans*
Site: Pigeon Key, FL
Magnification: 54X

Other species of bryozoa from warmer waters—

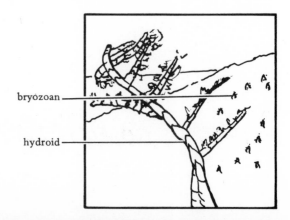

bryozoan

hydroid

Plate 6-61. Bryozoan *Membranipora tuberculata* encrusting gulfweed *Sargassum natans*
Site: Pigeon Key, FL
Magnification: 117✕

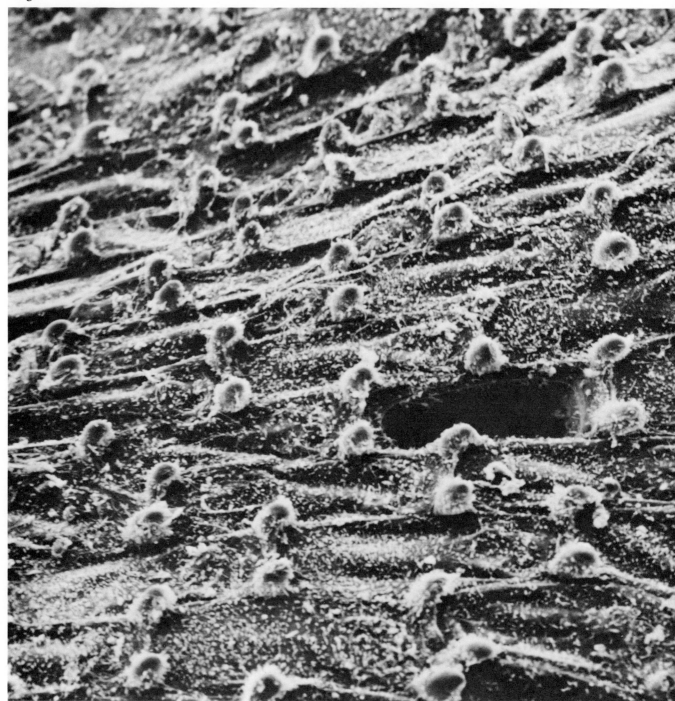

—are apparently free of the larger ciliated protozoa—

—and are colonized by bacteria, stalked diatoms, and the smaller protozoa—

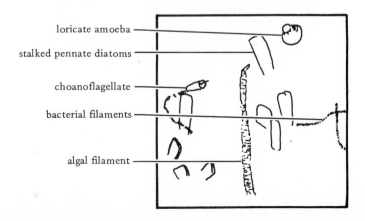

loricate amoeba

stalked pennate diatoms

choanoflagellate

bacterial filaments

algal filament

Plate 6-63. Plumularid hydroid on *Sargassum natans*
Site: Pigeon Key, FL
Magnification: 242X

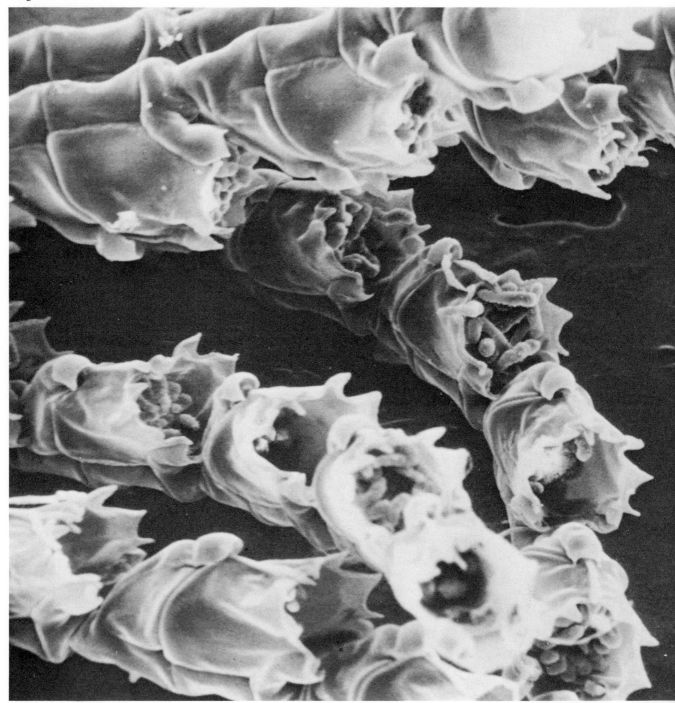

—while the adjacent hydroid is apparently free of microorganisms.

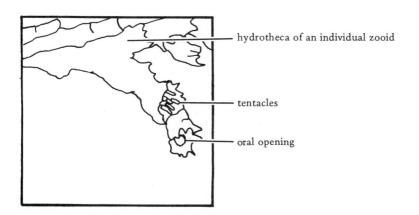

hydrotheca of an individual zooid

tentacles

oral opening

Plate 6-64. A nereid worm
Site: Narragansett Bay, RI
Magnification: 62X

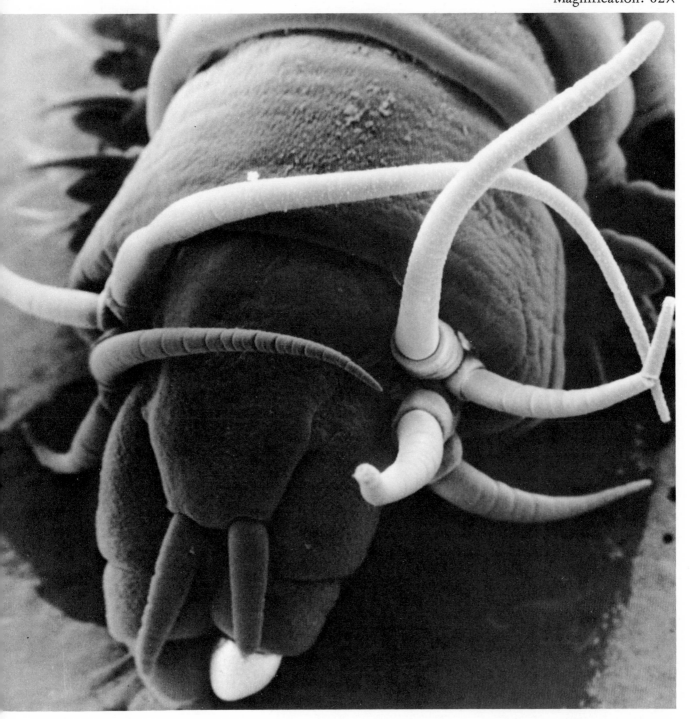

Other animals are also relatively free of microorganisms.

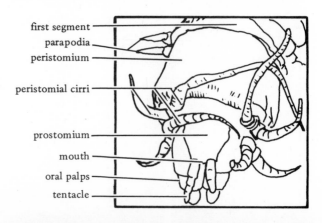

first segment
parapodia
peristomium

peristomial cirri

prostomium
mouth
oral palps
tentacle

Plate 6-65. Dorsal surface of worm
Site: Narragansett Bay, RI
Magnification: 3,817×

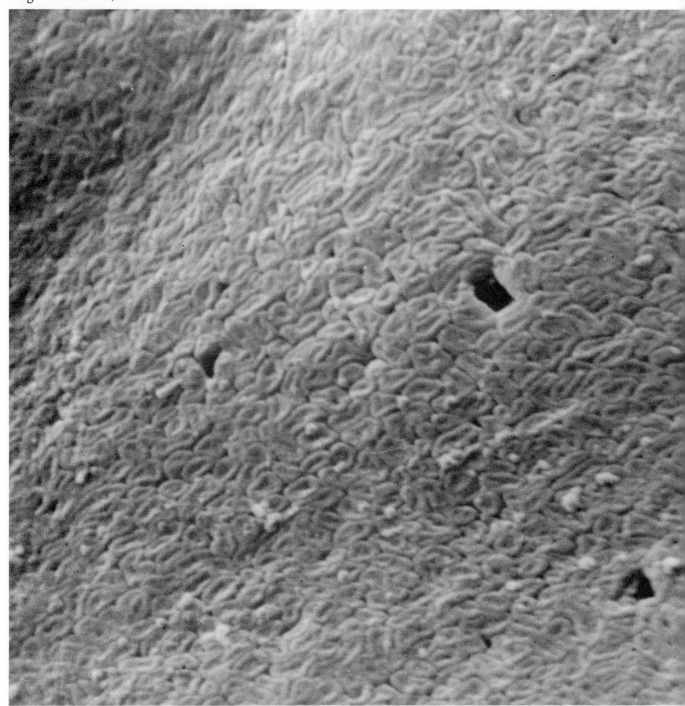

The textured skin with pores harbors only an occasional bacterial form.

This star-like creature containing one central cloacal opening and multiple oral openings—

Plate 6-67. The colonial tunicate *Botryllus schlosseri*
Site: Narragansett Bay, RI
Magnification: 140X

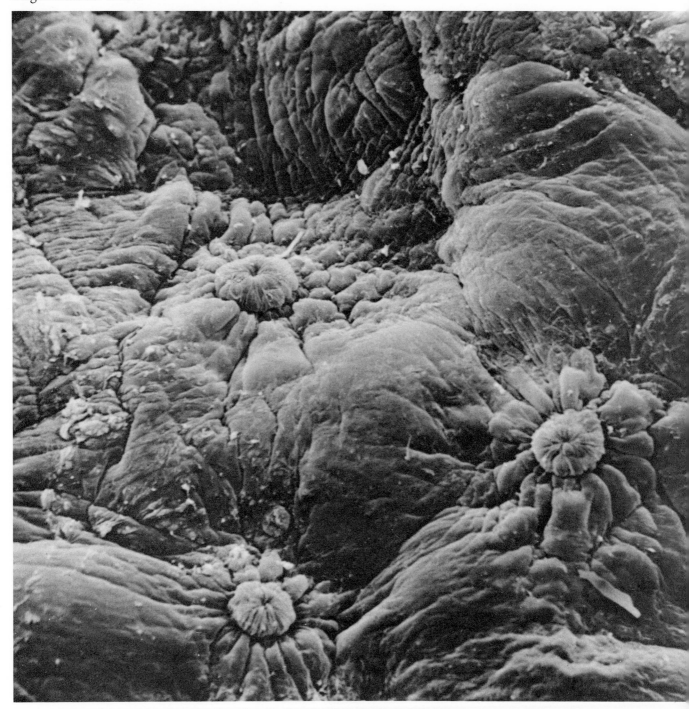

—*is another example of an animal with a relatively clean exterior surface.*

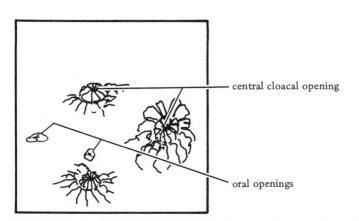

central cloacal opening

oral openings

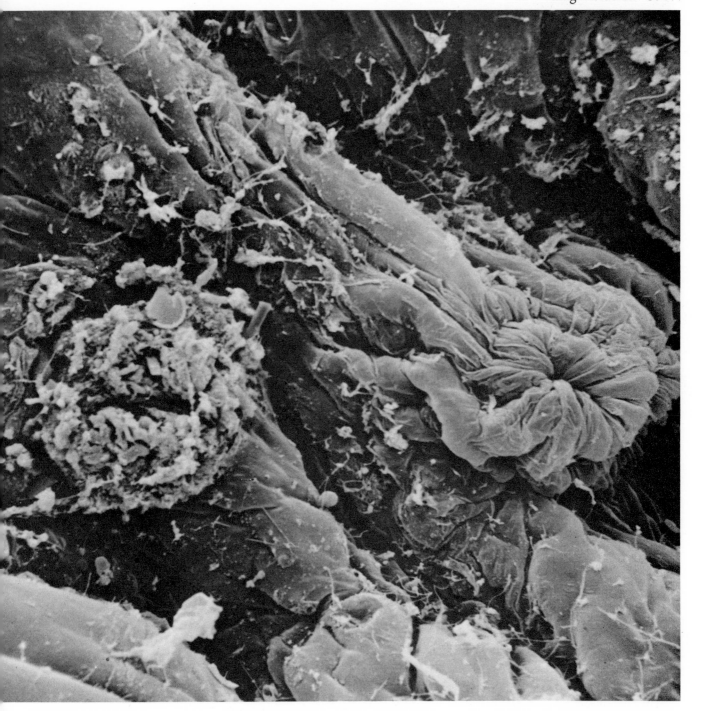

Ingested material forced out of the oral opening during fixation—

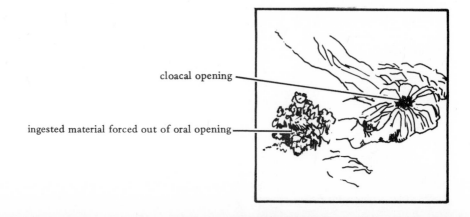

cloacal opening

ingested material forced out of oral opening

Plate 6-69. Ingesta of tunicate
Site: Narragansett Bay, RI
Magnification: 4,456×

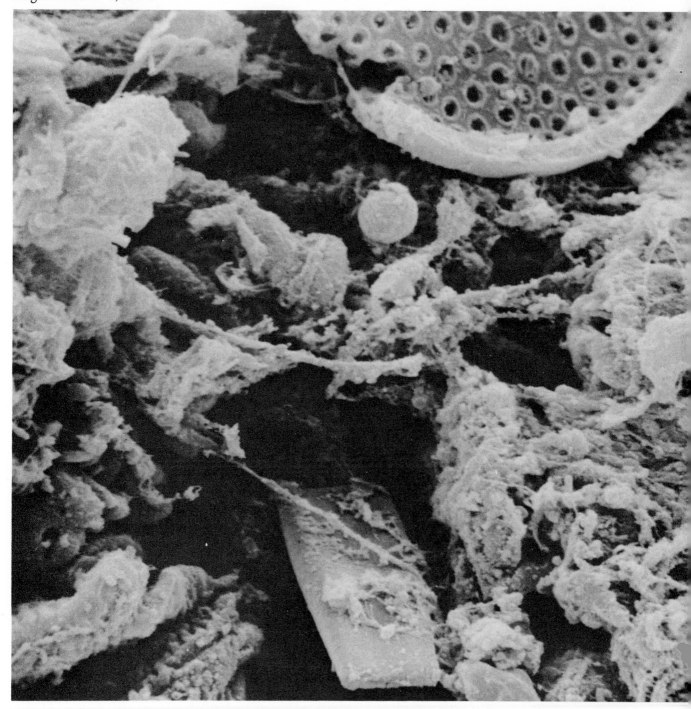

—*reveals diatom fragments and bacteria-rich organic debris.*

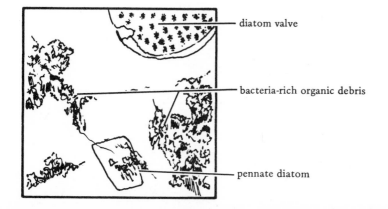

diatom valve

bacteria-rich organic debris

pennate diatom

The hydroid Obelia *is relatively free of microorganisms, while the area it shelters from the water currents—*

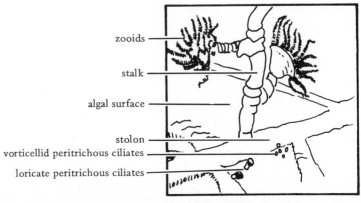

zooids

stalk

algal surface

stolon

vorticellid peritrichous ciliates

loricate peritrichous ciliates

Plate 6-71. Kelp surface sheltered by hydroid
Site: Narragansett Bay, RI
Magnification: 560X

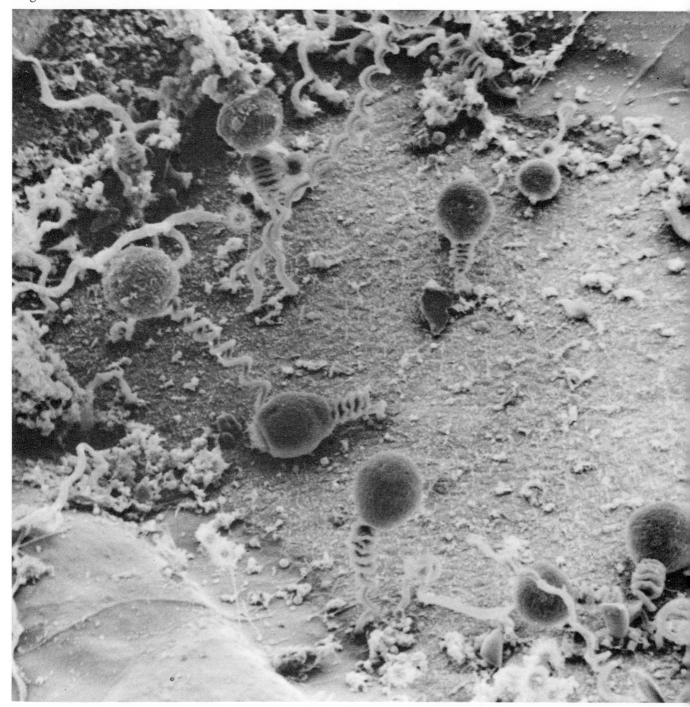

—supports protozoans attached to the plant surface—

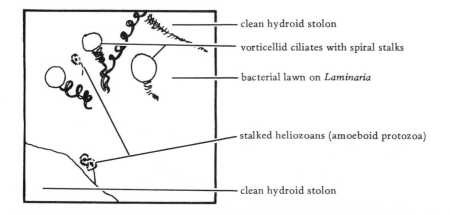

clean hydroid stolon

vorticellid ciliates with spiral stalks

bacterial lawn on *Laminaria*

stalked heliozoans (amoeboid protozoa)

clean hydroid stolon

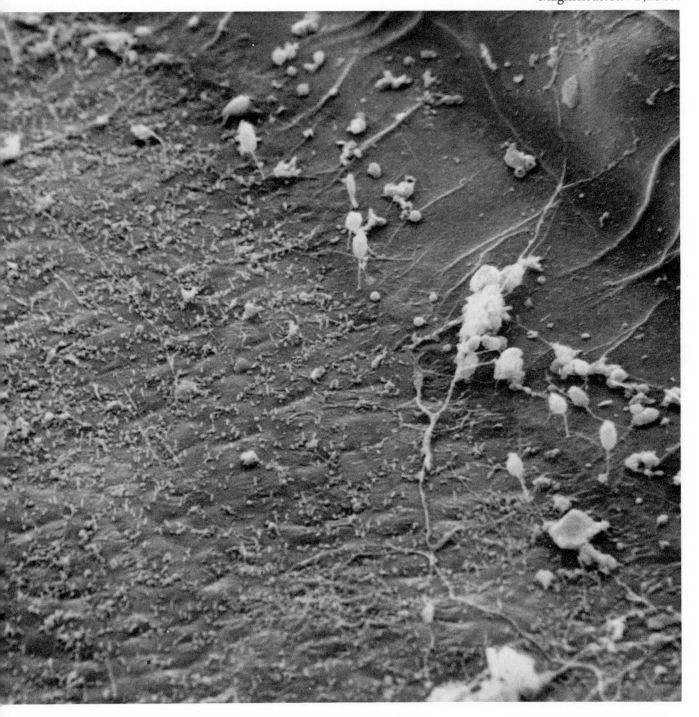

—as well as a rich bacterial lawn—

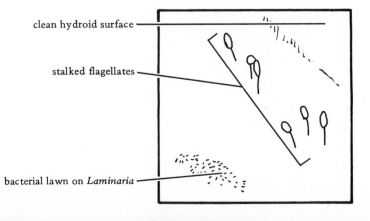

clean hydroid surface

stalked flagellates

bacterial lawn on *Laminaria*

Plate 6-73. Surface of hydroid stolon
Site: Narragansett Bay, RI
Magnification: 3,040X

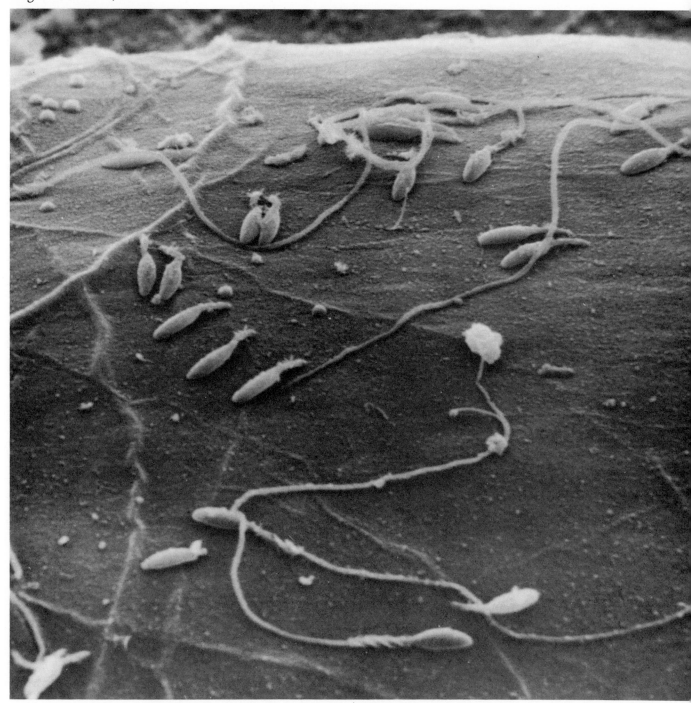

—while the relatively clean animal surface has discharged hydroid nematocysts (stinging cells) attached which can be confused with stalked bacteria by even an experienced observer.

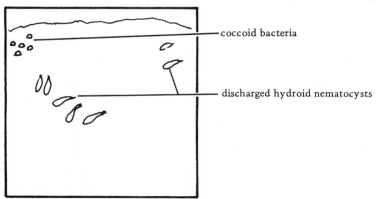

coccoid bacteria

discharged hydroid nematocysts

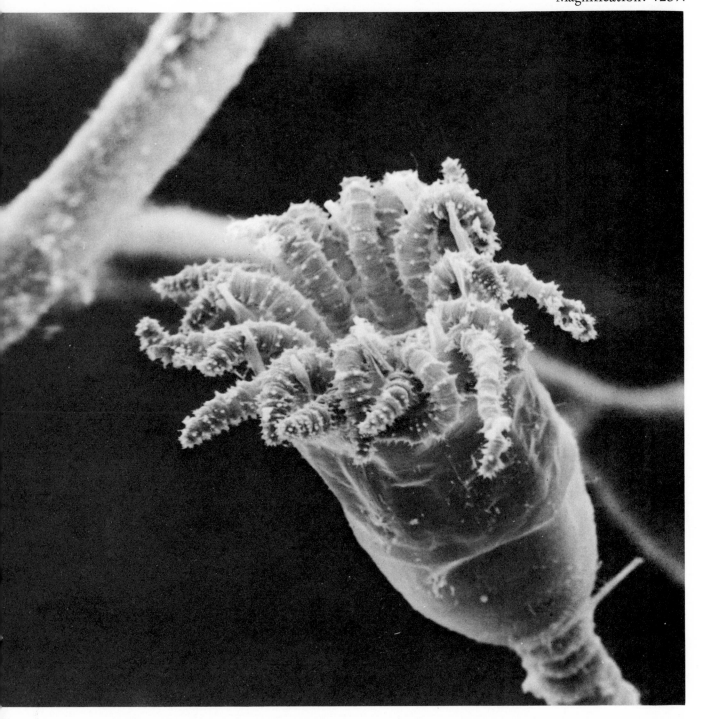

Other hydroid surfaces such as these warm water forms are also quite free of microorganisms—

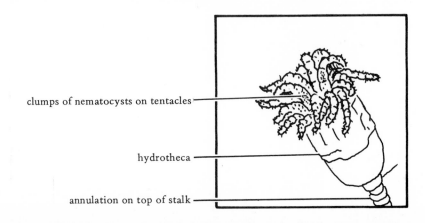

clumps of nematocysts on tentacles

hydrotheca

annulation on top of stalk

Plate 6-75. Stalk of campanularid hydroid
Site: North of Bermuda
Magnification: 2,840×

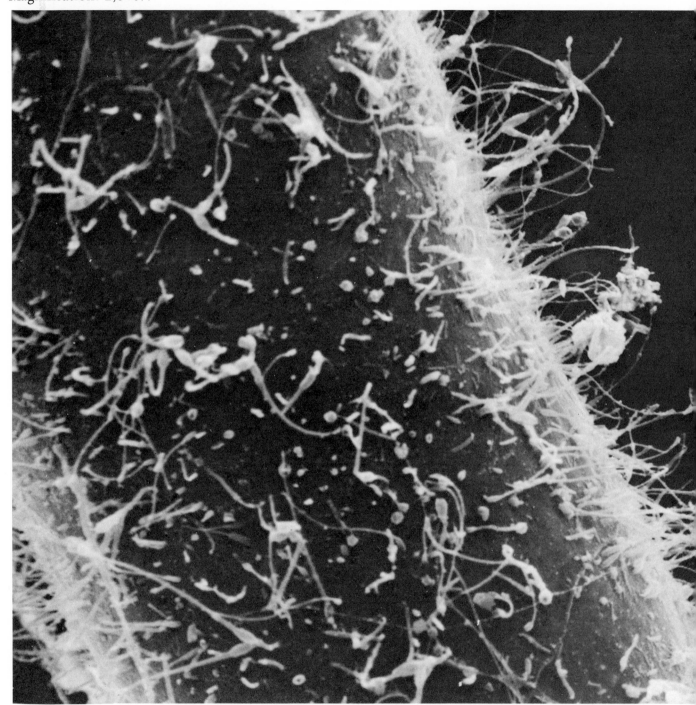

—except for the stalks, which can be heavily colonized with bacterial filaments.

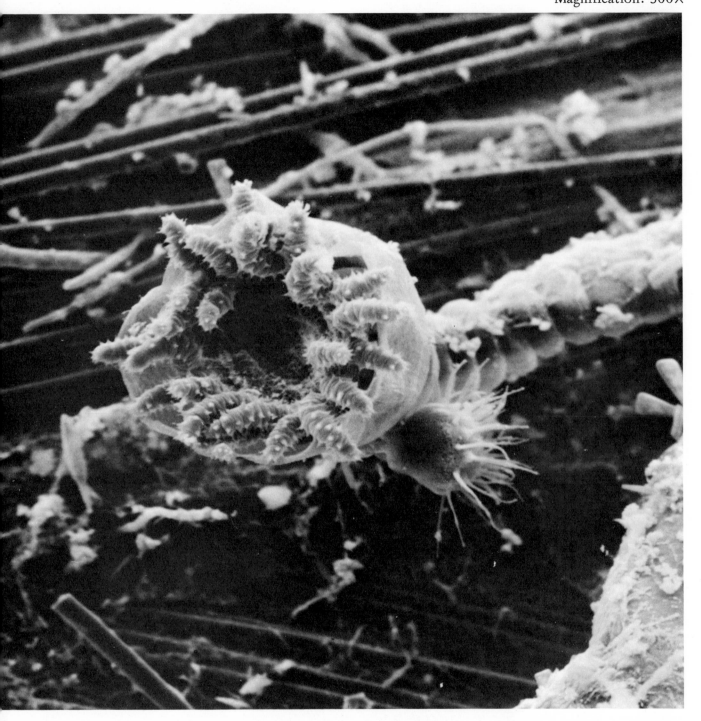

The aciliated stage of this ciliated protozoan, attached to the hydroid—

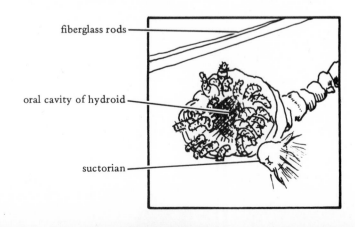

fiberglass rods

oral cavity of hydroid

suctorian

Plate 6-77. The suctorian surface
Site: North of Bermuda
Magnification: 2,460×

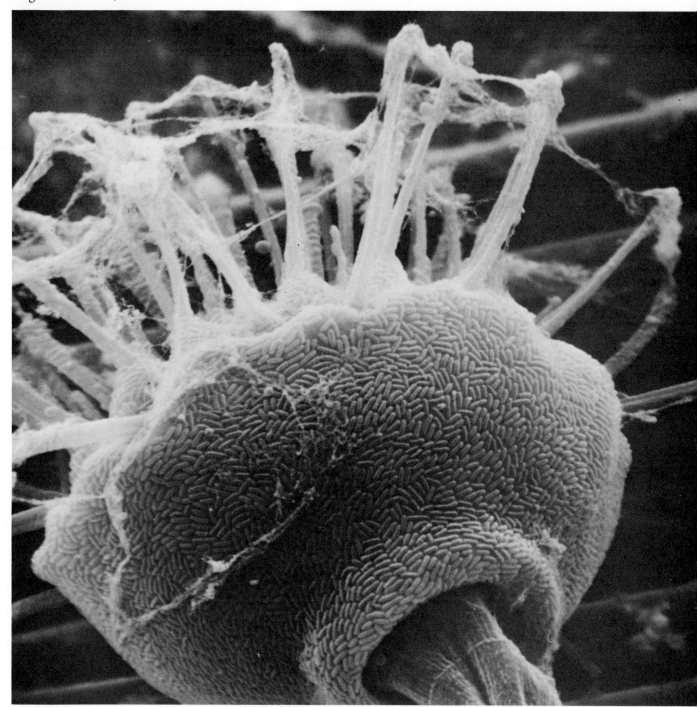

—is completely covered with bacteria similar to those on the vorticellid ciliate in Plate 6-51.

part seven

MICROBIAL DETERIORATION

This oak slat lodged on the sea floor has been perforated by boring isopods (limnoria) and molluscs (teredos) and softened by fungal attack.

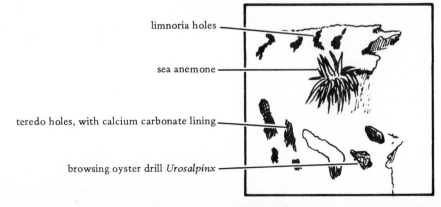

limnoria holes

sea anemone

teredo holes, with calcium carbonate lining

browsing oyster drill *Urosalpinx*

microbial deterioration

Organic materials such as wood, leather, hemp, and cotton soon deteriorate when exposed to the sea if precautions are not taken to inhibit their biodeterioration. Garbage, human wastes, and other organic debris have been assumed to undergo rapid attack in the sea and to be incorporated into the food chain. The sea has traditionally been used as a depository for man's refuse. However, in recent years we are realizing the finite limits of our planet and the delicate balance of the conditions required for biodeterioration processes to occur in the sea. The plastic age that has crept upon us during the last few decades has left us with a new flotsam and jetsam. Because microorganisms are not equipped with the enzymes necessary to attack such substances, these synthetic materials are accumulating at an alarming rate, not only at the sea's edge but throughout the water column (Carpenter and Smith, 1972; Colton, Knapp, and Burns, 1974). Moreover, they are even getting into the food chain and interfering with natural processes in the sea.

The paradoxical recovery of an intact and unspoiled lunch some 10 months after it had been accidentally submerged with the research submarine *ALVIN* at a depth of 1,540 meters in the Atlantic Ocean (Jannasch et al., 1971) dramatically showed our ignorance of biodeterioration in the deep sea. Subsequent studies on surface-inoculated samples (Jannasch et al., 1971) and in situ inoculated samples (Jannasch and Wirsen, 1973) have shown that bacterial deterioration per se is appreciably slowed by the cold temperature and hydrostatic pressure in the deep aphotic zone of the sea. The hydrostatic pressure of even 50 to 100 meters (5 to 10 atmospheres of pressure) seems to appreciably slow bacterial activities (Wirsen and Jannasch, 1974).

However, biodeterioration in the sea is not a single type of organic process but a series of events carried out by a sequence of macro- and microorganisms. A prerequisite for many microbial processes seems to be a prior or simultaneous attack by animals that chew holes through the interior of the material, permitting both microorganisms and oxygenated water to enter the material. Any process which interferes with this inhibits the process of biodeterioration. The remnant of an oak lobster pot shown in Plate 7-1 is an example of biodeterioration carried out by microorganisms and animals working in concert. A preliminary attack on the more readily utilized organic substances by bacteria leaves the more resistant cellulosic materials. Fungi which can attack these materials readily colonize the surface and penetrate the wood through its vessels and pores (Brooks, Goos, and Sieburth, 1972). The cavities left by the boring isopods (limnoria) greatly facilitate fungal development. The fungi in turn attract the shipworms (teredos), which are molluscs with a shell adapted for boring burrows through immersed wood, leaving behind a labyrinth of non-connecting tunnels which weaken the wood. The final soft rot stage is accomplished by fungi and is characterized by the appearance of distinctive fruiting bodies. The biology of the biodeterioration of wood is summarized in symposia proceedings edited by Ray (1959) and Jones and Eltringham (1971).

Another example of animal acceleration of biodeterioration is the kelp fly, which lays its eggs on seaweed cast upon the shore. The developing larvae penetrate the seaweed, permitting a more rapid development of yeasts and bacteria which further decompose the seaweed (Bunt, 1955). The outer part of such seaweeds seems to be colonized and perhaps decomposed by pennate diatoms of the genera *Cocconeis* and *Nitzschia* which have a marked heterotrophic activity. In the sea a variety of snails and amphipods perforate seaweed tissue and presumably hasten its microbial decomposition. This process of animal—microbial biodeterioration seems to be quite labile, being inhibited by conditions which deny the access of oxygenated water and animals. Such conditions may have existed in the well preserved *ALVIN* box lunch (Sieburth and Dietz, 1974). When the delicate requirements of biodeterioration are met, it is an effective mechanism for the recycling of organic carbon and other materials. However, the present method of dumping wastes in bales or in accumulations minimizes exposure and animal penetration and leads to preservation rather than to biodeterioration.

The biodeterioration of materials such as fat and flesh is more difficult to observe than that of plant materials. Dead and injured animal tissue is soon perforated or trimmed by carnivorous and omnivorous amphipods and isopods which exist even in the deep sea (Hessler, Issacs, and Mills, 1972). An example of the deterioration of animal tissues is shown in the series of micrographs documenting the microbial sequence of tail rot in chinook salmon smolts reared in a tank where the tails could be easily abraded. The normal skin surface with its patterned mucoid material is apparently very resistant to bacterial colonization. Once this protective mechanism is bypassed, as through abrasion, bacteria quickly develop on the exposed flesh. As oxygen is depleted microzonally, spirochetes also become prevalent. Bacteria soon decompose the fleshy material, exposing the tail cartilage. At this point the fungus *Saprolegnia* turns the tail stub into a highly visible fuzz of fungal mycelia unless the primary bacterial erosion is curtailed with antibiotic treatment.

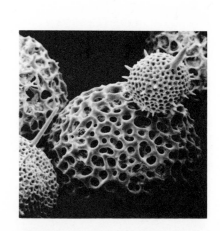

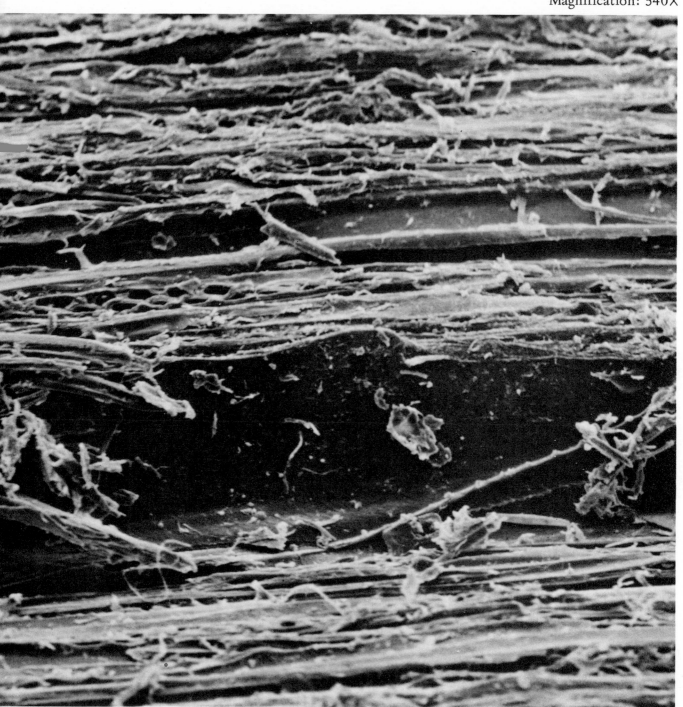

A newly submerged wood surface—

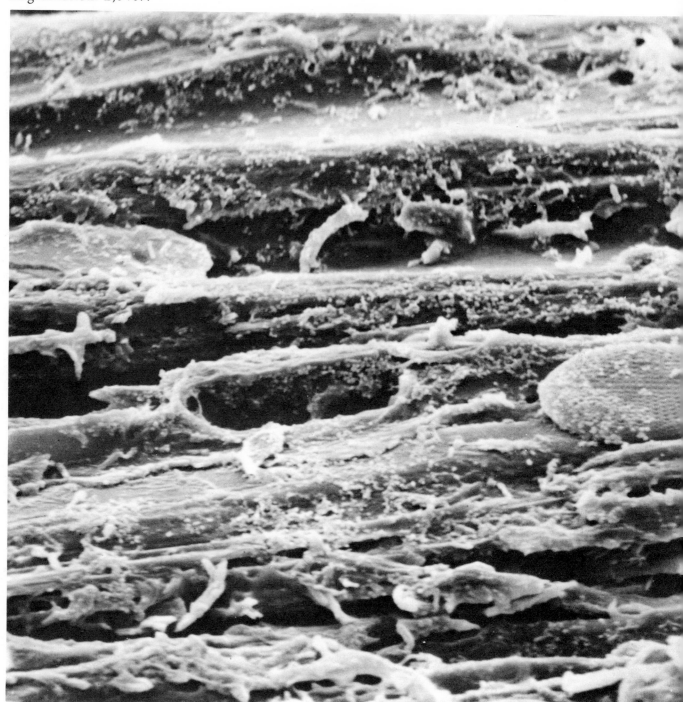

—*apparently provides nutrients for the rapid formation of a bacterial slime.*

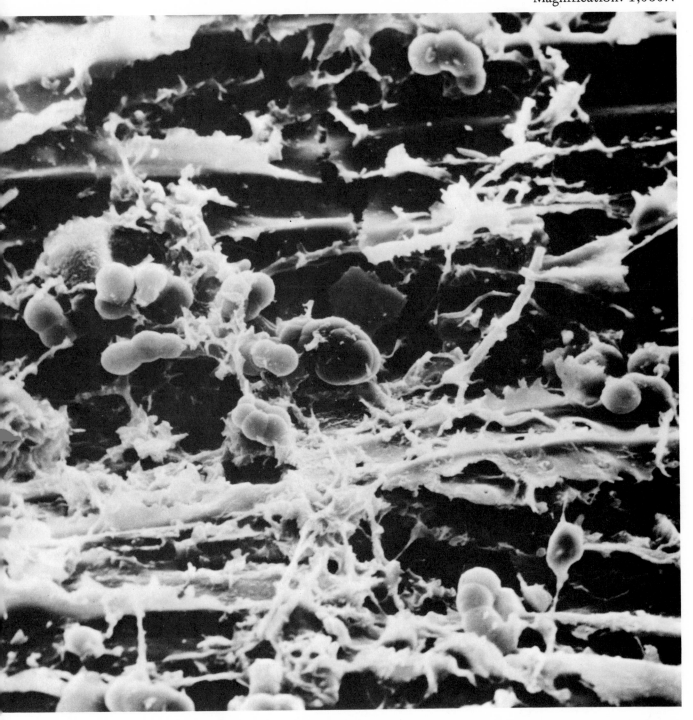

The fungi, distinguished by their characteristic spores, soon develop on the wood surface—

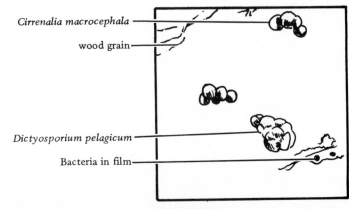

Cirrenalia macrocephala

wood grain

Dictyosporium pelagicum

Bacteria in film

Plate 7-5. Mold spores on wood surface
Site: Point Judith, RI
Magnification: 800X

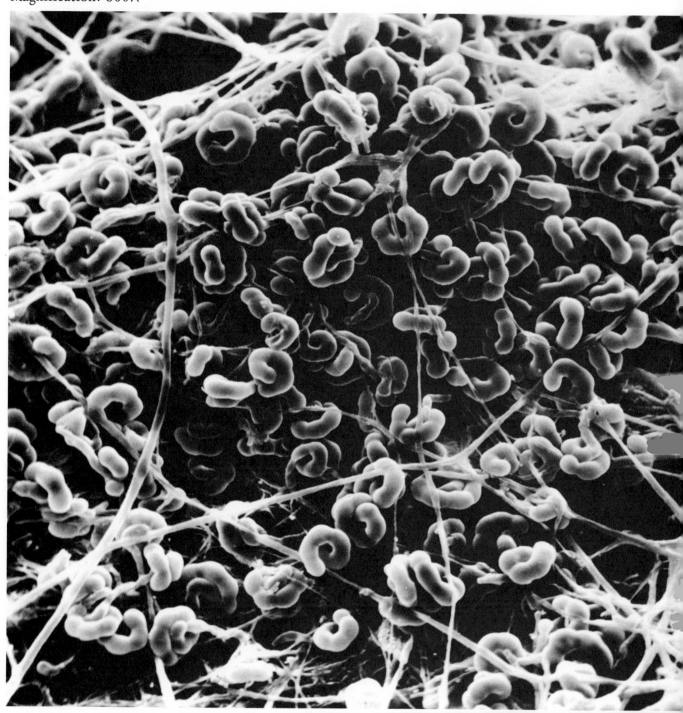

—*where they can form tremendous densities under ideal conditions, as shown on this wood fragment incubated in the laboratory.*

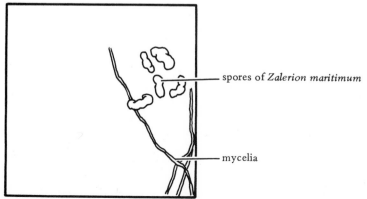

spores of *Zalerion maritimum*

mycelia

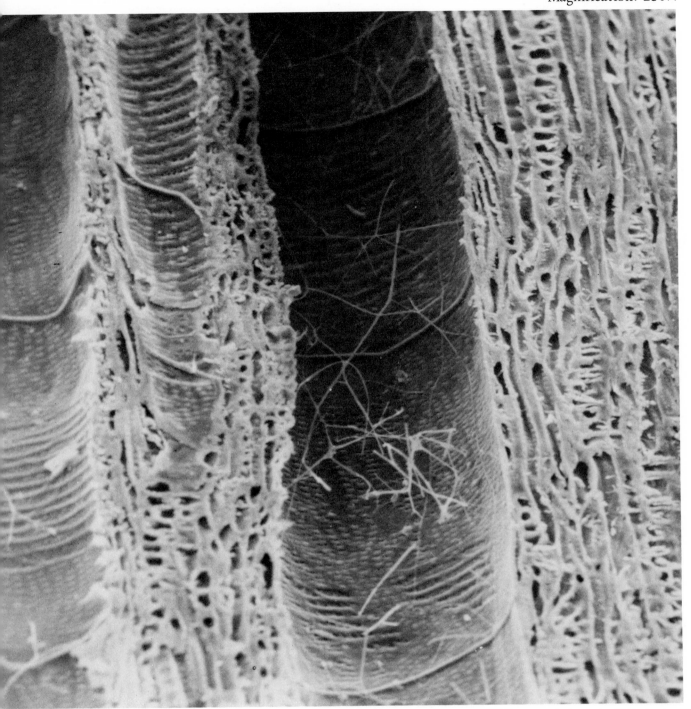

The fungal mycelia penetrate through the pores to the vessels—

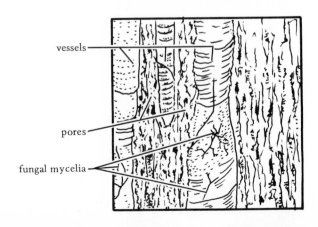

vessels

pores

fungal mycelia

Plate 7-7. Fungal mycelium in amphipod cavity
Site: Point Judith, RI
Magnification: 110✕

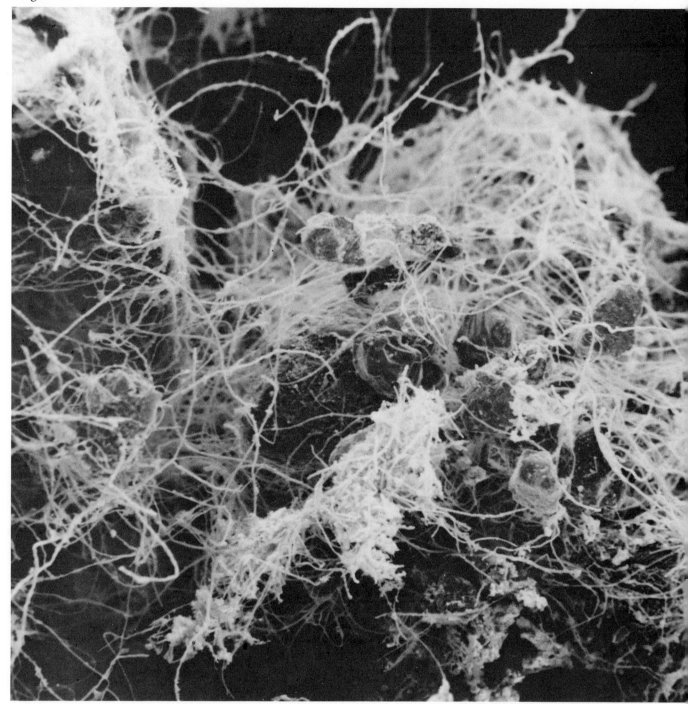

—and into cavities produced by limnoria which provide the wood-destroying molds with oxygen and a route to spread through the wood.

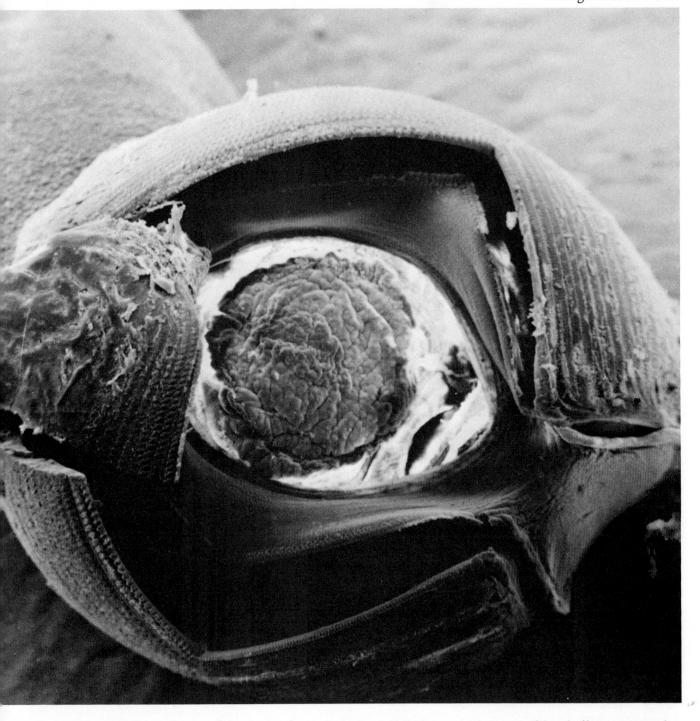

The teredo, which produces a labyrinth of wood-weakening burrows with its saw-like shell, is attracted to decaying wood by the fungi—

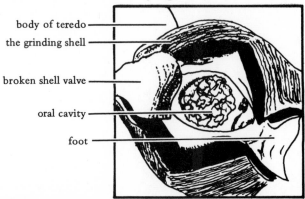

body of teredo
the grinding shell
broken shell valve
oral cavity
foot

Plate 7-9. Rotted wood containing fungal fruiting bodies (perithecia)
Site: Point Judith, RI
Magnification: 100X

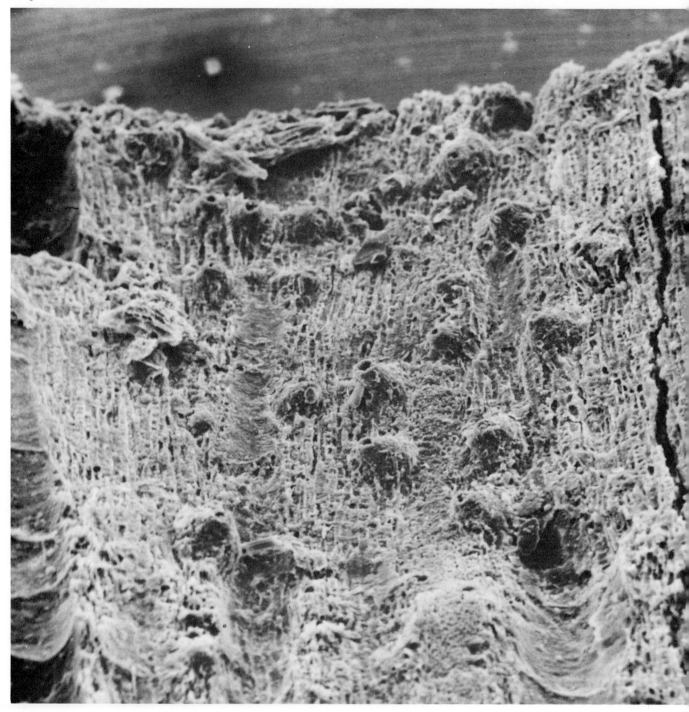

—which produce fruiting bodies during an advanced state of decay.

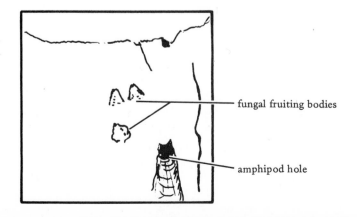

fungal fruiting bodies

amphipod hole

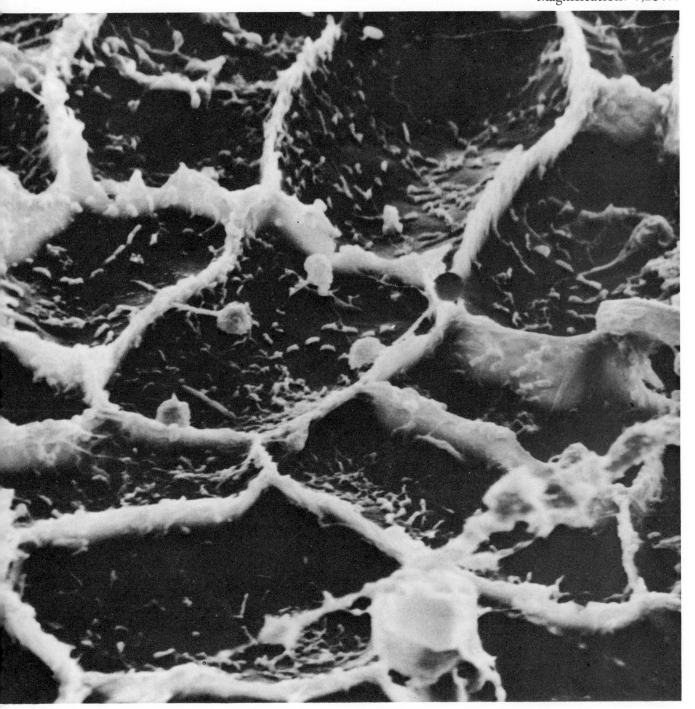

Terrestrial plant materials are readily colonized by bacteria such as are seen on the underside of this leaf—

Plate 7-11. Decaying mangrove leaf
Site: Molasses Key, FL
Magnification: 2,150×

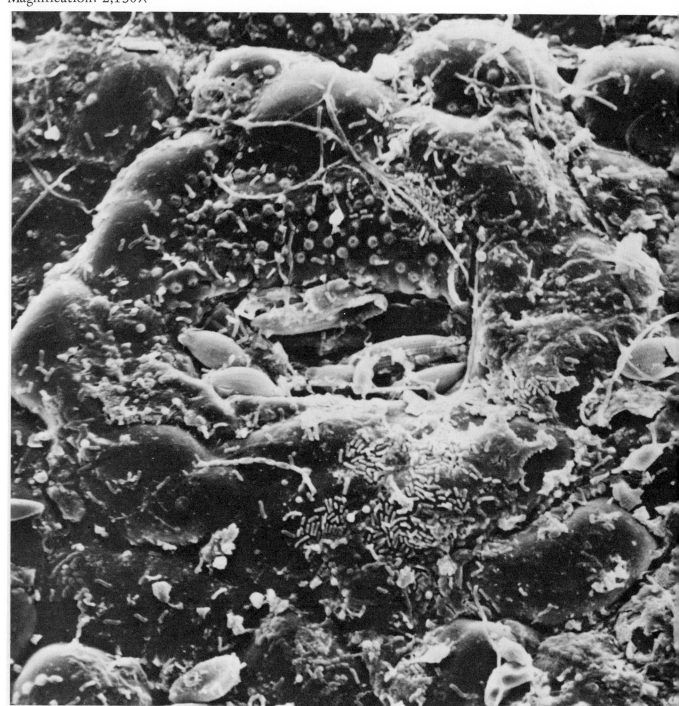

—and the area around the stomata of this tropical one.

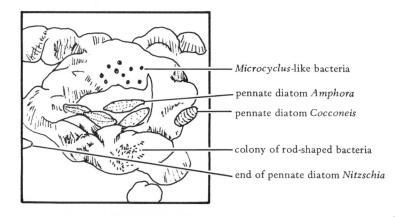

Microcyclus-like bacteria

pennate diatom *Amphora*

pennate diatom *Cocconeis*

colony of rod-shaped bacteria

end of pennate diatom *Nitzschia*

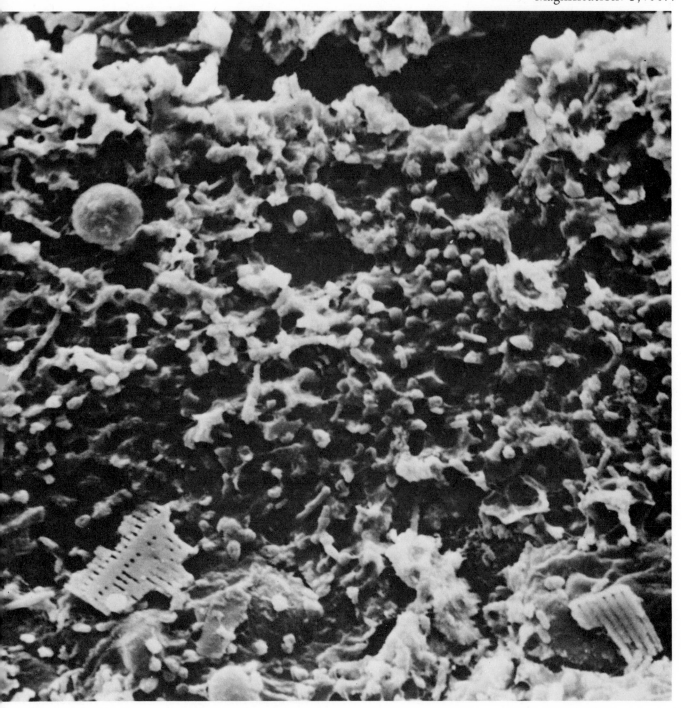

Intense bacterial development on the surface of an unidentified twig—

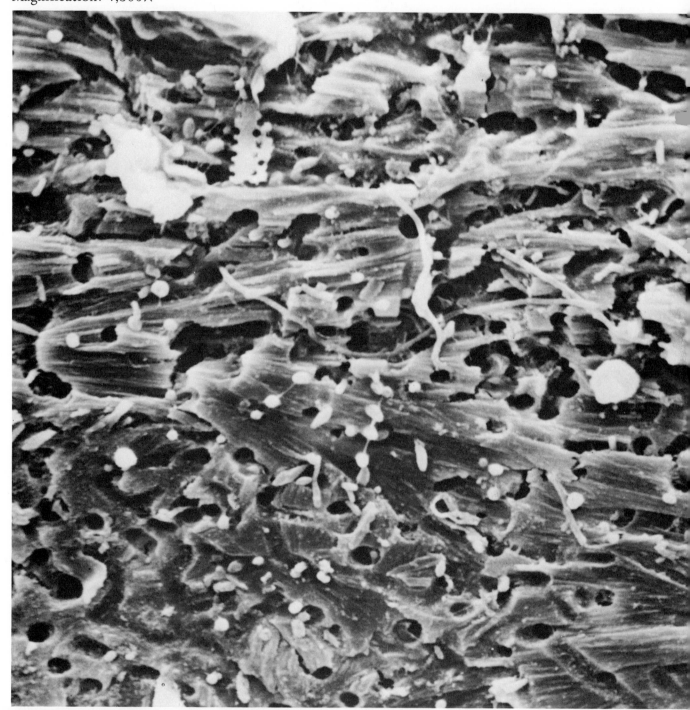

—has apparently left bacteria-sized perforations in the epithelial layer.

Plate 7-14. Dialysis membrane (regenerated cellulose film) submerged for one day
Site: Narragansett Bay, RI
Magnification: 4,290X

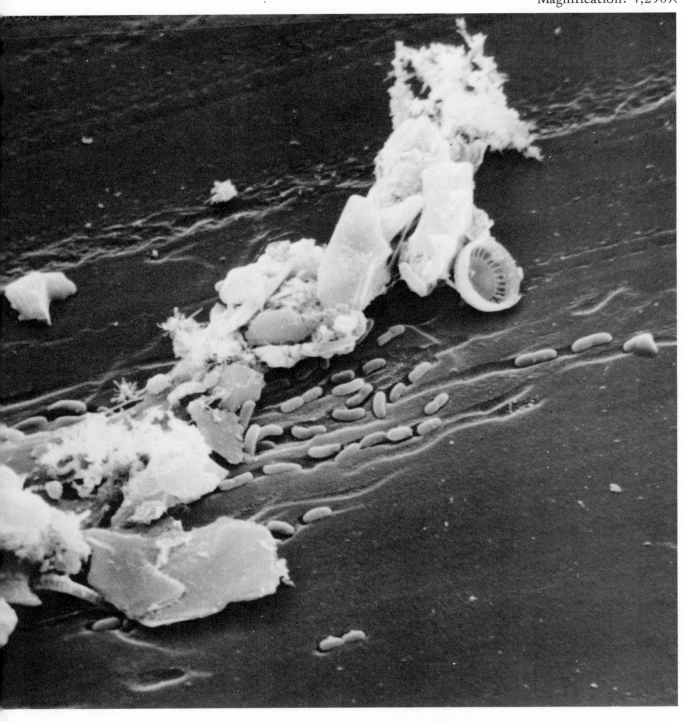

Bacteria attacking cellulose are apparent within one day—

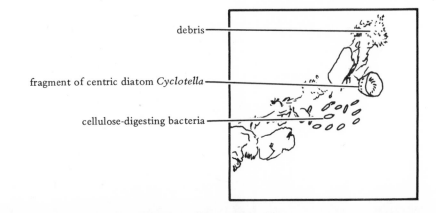

debris

fragment of centric diatom *Cyclotella*

cellulose-digesting bacteria

Plate 7-15. Regenerated cellulose film submerged for three days
Site: Narragansett Bay, RI
Magnification: 2,630X

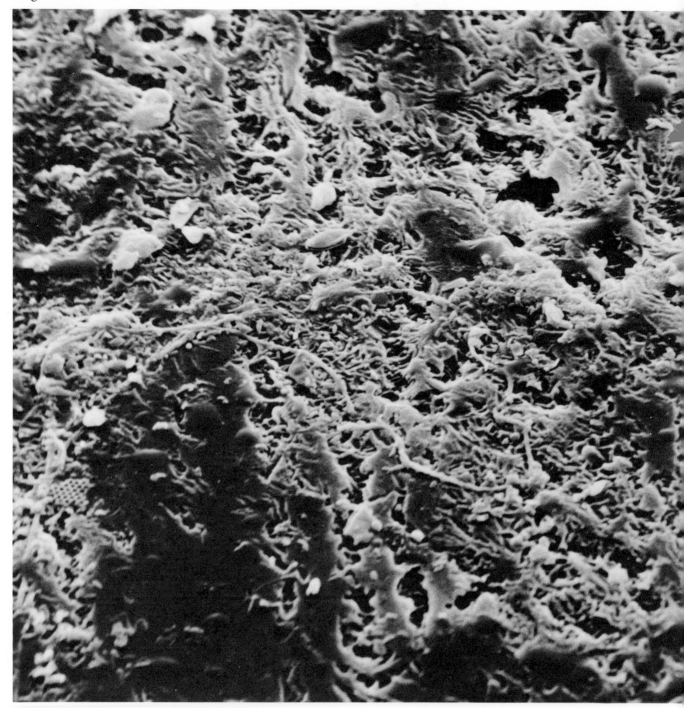

—and develop populations which extensively scar and perforate the surface by the end of three days.

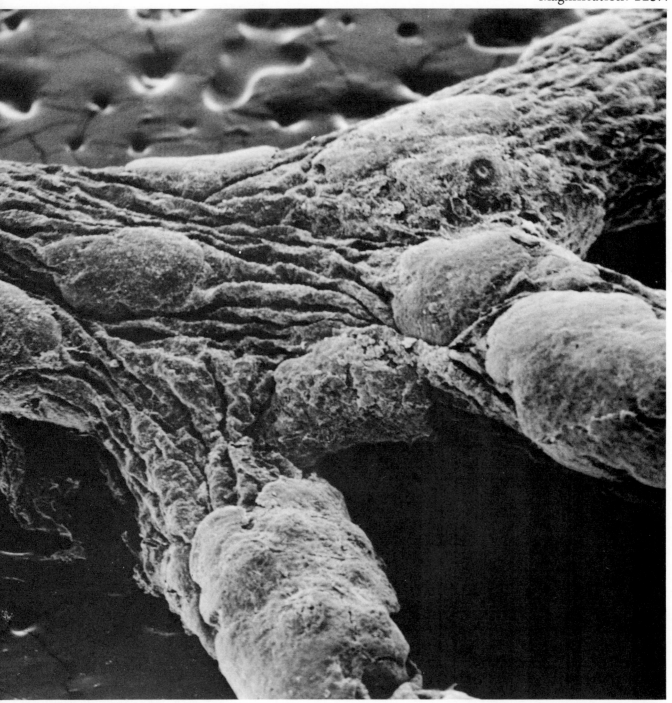

The surface of decaying seaweed—

Plate 7-17. Microorganisms on surface of decaying seaweed
Site: Pigeon Key, FL
Magnification: 1,050X

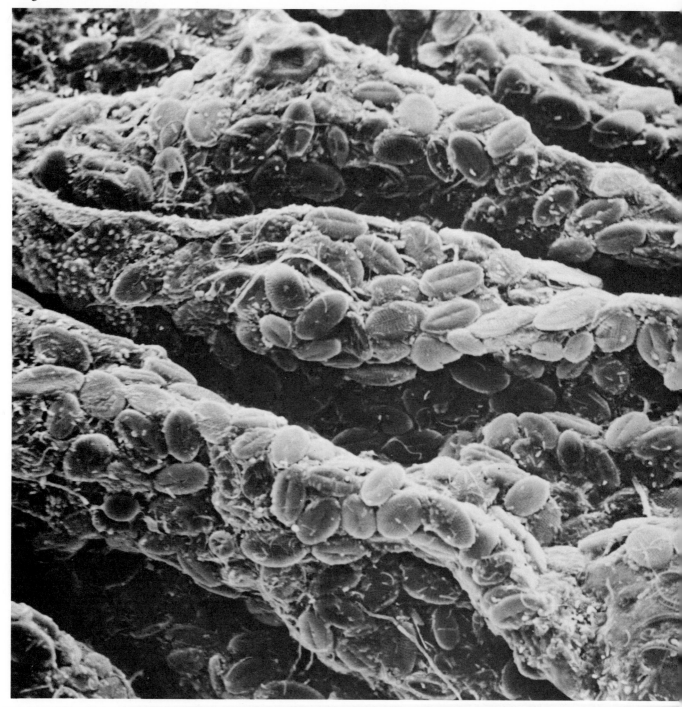

—is covered by a mixture of diatoms and bacteria—

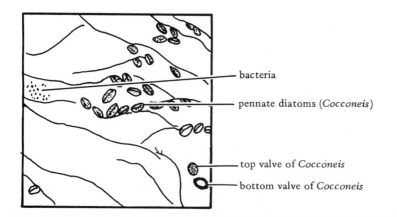

bacteria

pennate diatoms (*Cocconeis*)

top valve of *Cocconeis*

bottom valve of *Cocconeis*

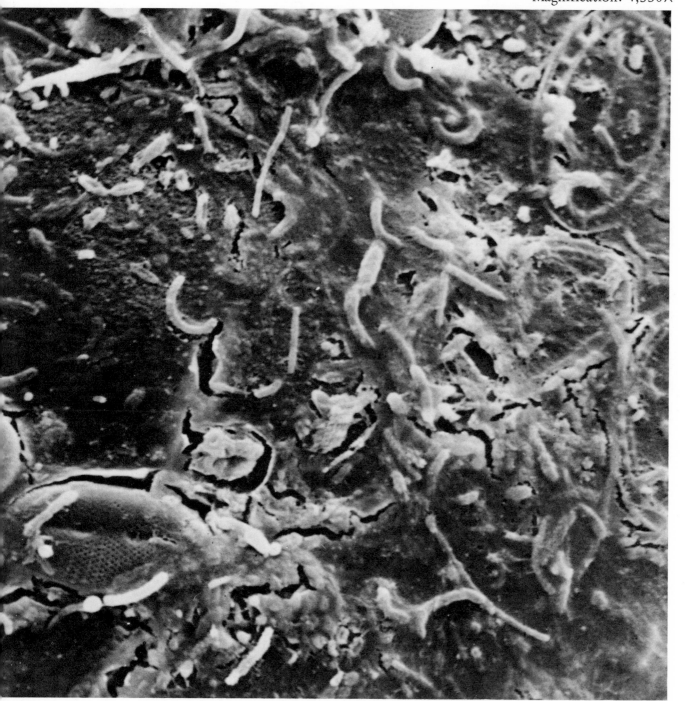

—embedded within a layer of slime.

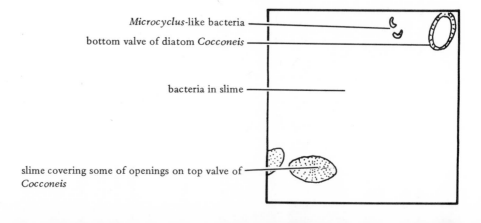

Microcyclus-like bacteria

bottom valve of diatom *Cocconeis*

bacteria in slime

slime covering some of openings on top valve of *Cocconeis*

Plate 7-19. Chinook salmon smolts *Oncorhynchus tshawytscha*
Source: University of Rhode Island Aquaculture Project
Magnification: 2X

The form and transparency of the tail changes during the progressive stages of tail rot.

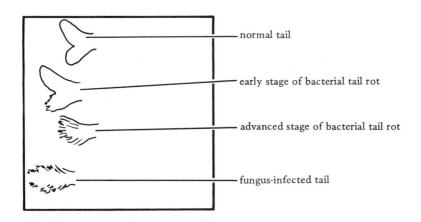

normal tail

early stage of bacterial tail rot

advanced stage of bacterial tail rot

fungus-infected tail

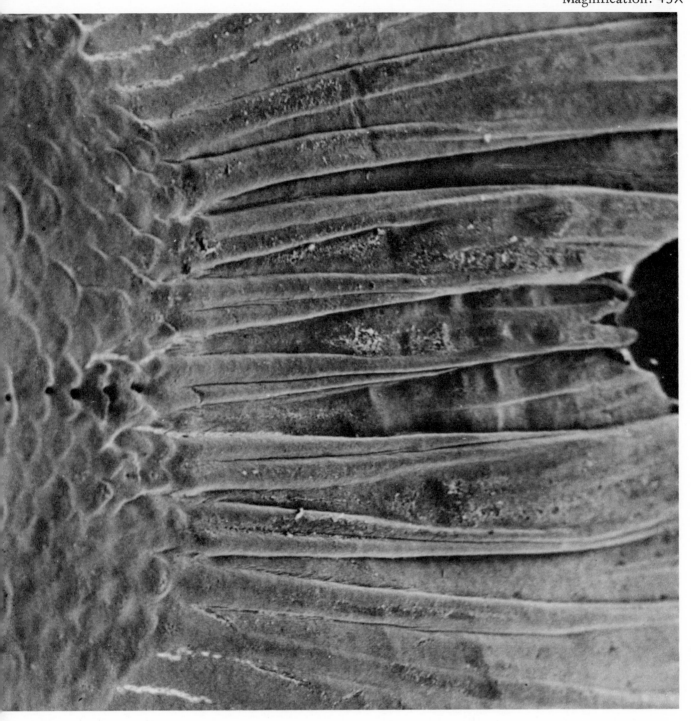

The undamaged tail—

Plate 7-21. Patterned surface layer on skin of a salmon tail
Source: University of Rhode Island Aquaculture Project
Magnification: 4,700X

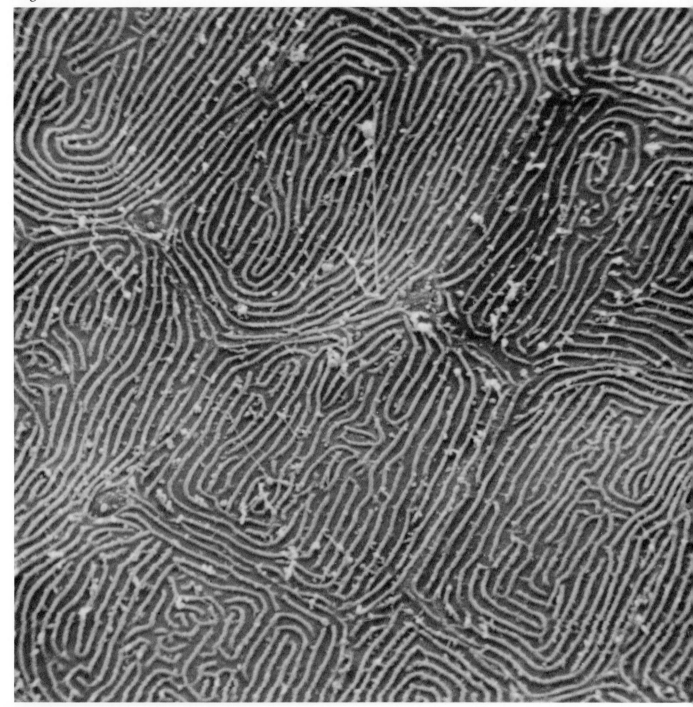

—is covered by a thumbprint-like epithelium free of microorganisms.

Plate 7-22. Decaying salmon tail with exposed cartilage
Source: University of Rhode Island Aquaculture Project
Magnification: 60X

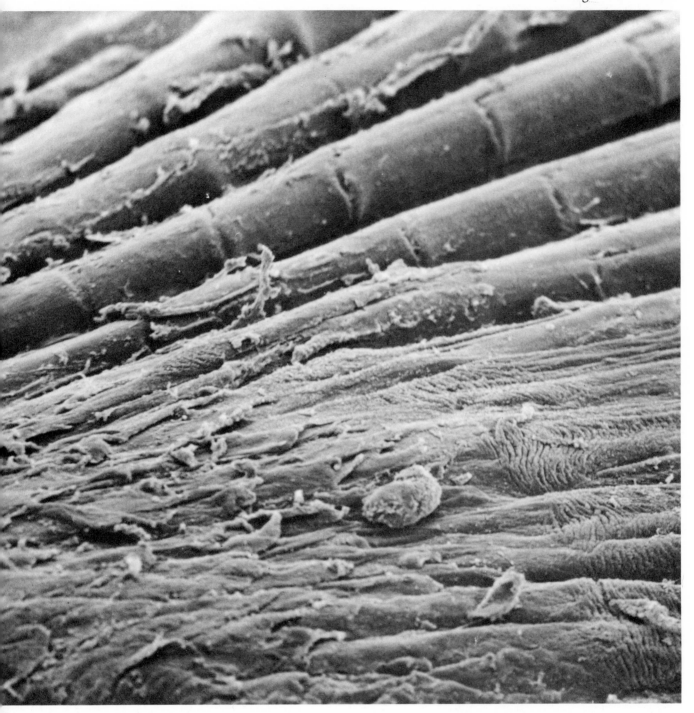

In fish that have been affected, the tail cartilage is exposed by the decomposition of the flesh—

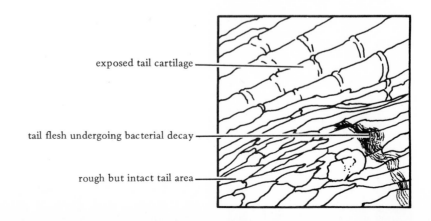

exposed tail cartilage ———

tail flesh undergoing bacterial decay ———

rough but intact tail area ———

Plate 7-23. Decaying flesh on salmon tail
Source: University of Rhode Island Aquaculture Project
Magnification: 4,630×

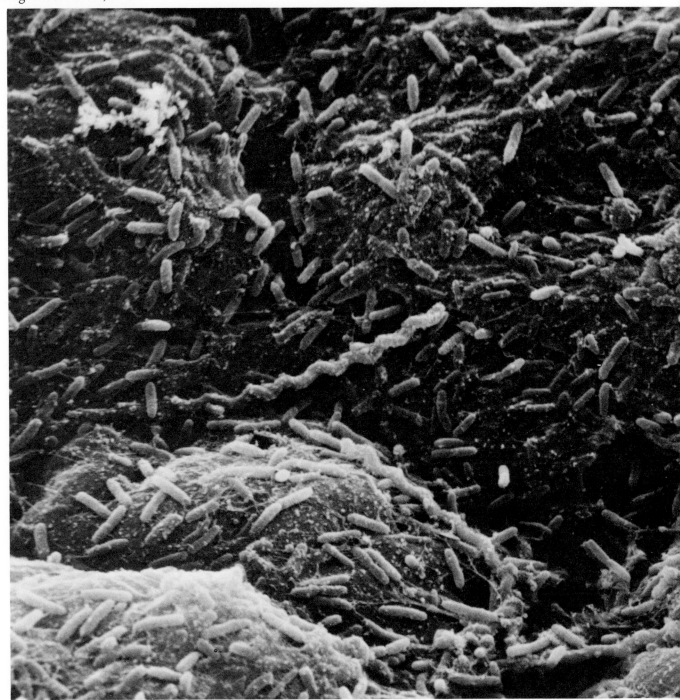

—*by bacterial rods and spirochetes.*

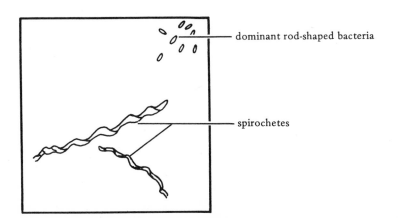

dominant rod-shaped bacteria

spirochetes

The terminal stage of tail rot is the appearance of a whitish fuzz caused by fungal filaments—

Plate 7-25. Mycelia and sporangia of the fungus *Saprolegnia*
Source: University of Rhode Island Aquaculture Project
Magnification: 530X

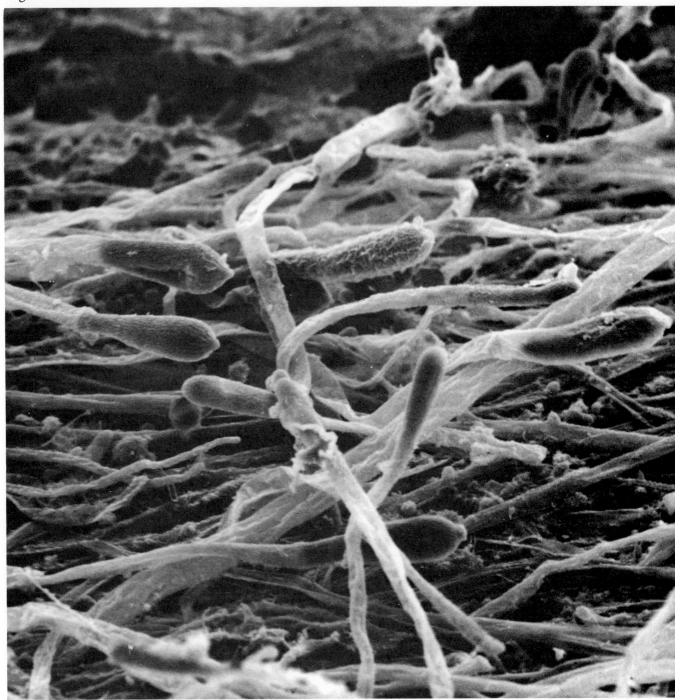

—identifiable as Saprolegnia *by the characteristic sporangia.*

part eight

SEA FLOOR

A trail left on the mud by this snail shows the plasticity of the organic debris and mineral fragments deposited on the sea floor in estuarine waters.

sea floor

The rocky crust of the earth is exposed at the turbulent edge of the sea. These rocks are colonized like other "inanimate" surfaces, the extent of microbial colonization depending upon the texture of the surface and the degree of turbulence, among other factors. A sandstone surface in an exposed and turbulent area where barnacles settle and thrive was found to support an algal mat which in turn supported the filamentous bacterium *Leucothrix mucor,* which has a special ability to attach to surfaces (Harold and Stanier, 1955; Pringsheim, 1957). Pebbles found in the surf zone are worn smooth by the abrasive action of the sand and are very lightly colonized, while those in deeper and protected waters support a richer microbiota of pennate diatoms, bacteria, and protozoa. The sea floor is covered by sands, silts, and clays arising from the erosion of rocks and minerals on land and from the siliceous and calcareous remains of microorganisms, plants, and animals that live in the sea and settle to the sea floor.

The microstructure of the mineral particles may dictate the availability of protected microhabitats where microorganisms can develop. Also important to microbial activity on the sea floor are the grain size and pore size of the sediments which control the type and amount of organic debris trapped by the sediment. A series of samples from the sand grains in the surf zone to the fine black mud in quieter waters shows this quite distinctly. The sand grains exposed to the surge of the sea are larger and more barren of microorganisms than subtidal sand grains that do not receive the direct action of the surf. The hollow or concave surfaces of sand grains are colonized by patches of pennate diatoms, blue-green algae, and bacteria. Abrasion apparently prevents the colonization of convex and exposed flat surfaces. The nature and distribution of microorganisms on sand grains has been reviewed and studied by Meadows and Anderson (1968). The mineral particles become finer with depth and in protected areas such as coves. In eutrophic areas organic debris becomes trapped to form rich muds that become anaerobic and produce sulfides which blacken them. The nature, chemistry, and microbiology of such sediments are well described in the monograph by Fenchel (1969).

Micrographs of these rich muds fail to show the protozoan and metazoan faunas that develop, but they do show the organic debris, fecal pellets, and amorphous bacterial flocs which feed the fauna. This organic matter is much more abundant than the fine mineral particles found in these sediments, but the rich bacterial flora in these flocs is largely masked by their own polysaccharide slime. The microbial flocs which provide the basic nutrition for "deposit-feeding" animals are in turn supported by organic debris such as that from the turtle grass *Thalassia testudinum* which supports a microbial sequence of bacteria, flagellates, and ciliates whose populations increase with decreasing size of the particles (Fenchel, 1970).

As water depth increases and light is attenuated below the level required by photosynthetic microorganisms, the heterotrophic bacterial flora in the sediments becomes more dependent upon the organic debris that settles through the water column to the sea floor. In the deep open ocean where the production of organic matter in the relatively thin euphotic zone is sparse compared to that of inshore and continental shelf waters, the incorporation of organic matter into the sediments is correspondingly sparse. In these deep areas (2,000 to 5,000 meters), the sediments are oozes formed by the calcareous and siliceous skeletons of particle-ingesting protozoans such as the Radiolaria and Foraminifera, and of photosynthetic microorganisms such as the diatoms and coccolithophorid flagellates. The amorphous organic matter present in these oozes is indistinguishable from the bacterial flocs found in the sediments of inshore waters, and bacteria can be demonstrated by cultural procedures (Morita and ZoBell, 1955). A few examples of such sediments and their "boneyards" give only the barest suggestion of the architectural beauty of the skeletons of these amazing microorganisms (Haeckel, 1887; Moore, 1954; Phleger, 1960). In the deeper abyssal plains the oozes give way to the red clays which are comparatively barren of protozoan skeletons, organic matter, and bacteria.

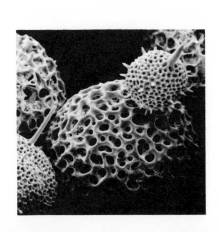

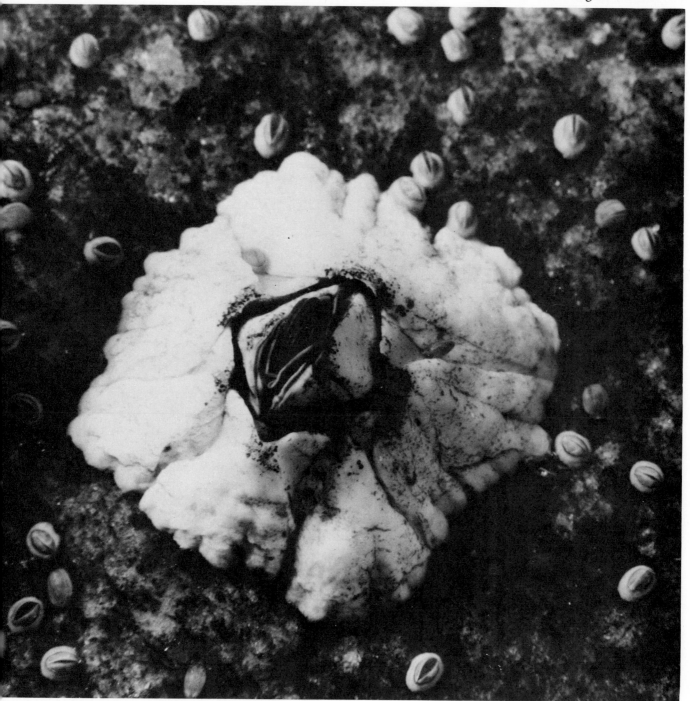

The sandstone rock is encrusted with an algal mat—

Plate 8-3. Microorganisms on an algal mat encrusting sandstone
Site: Narragansett Bay, RI
Magnification: 3,720×

—which becomes littered with debris and populated by assorted bacteria including the filamentous bacterium Leucothrix mucor.

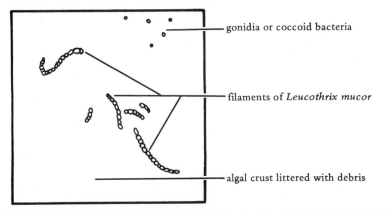

gonidia or coccoid bacteria

filaments of *Leucothrix mucor*

algal crust littered with debris

Plate 8-4. Pebble
Site: Narragansett Bay, RI
Magnification: 50X

Abrasion caused by tumbling of this pebble in the surf zone—

Plate 8-5. Pebble
Site: Narragansett Bay, RI
Magnification 500X

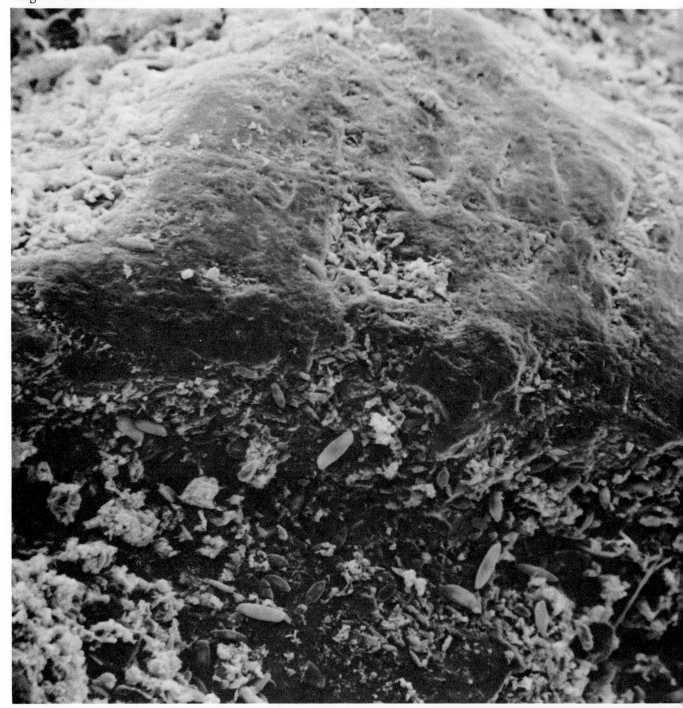

—has smoothed its exposed surfaces, leaving debris and microorganisms in the protected crevices—

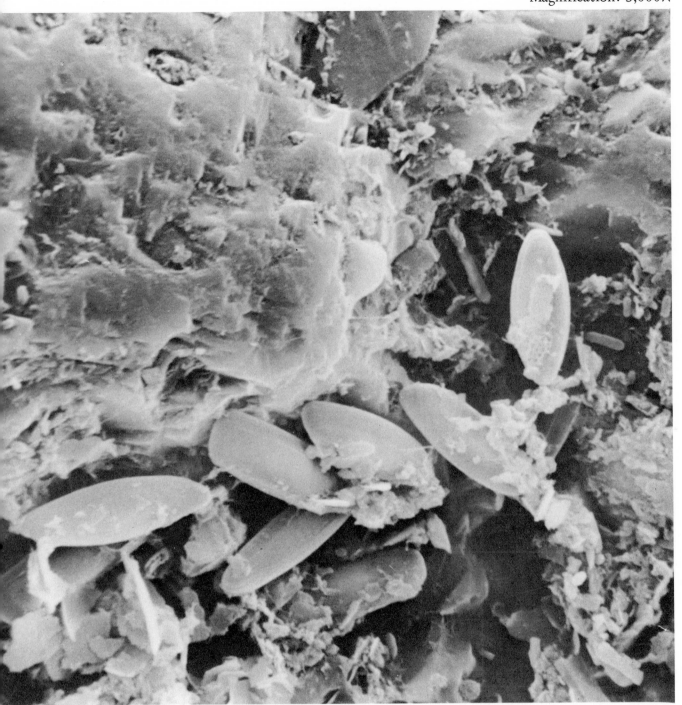

—consisting of pennate diatoms—

Plate 8-7. Microorganisms on pebble
Site: Narragansett Bay, RI
Magnification: 4,500×

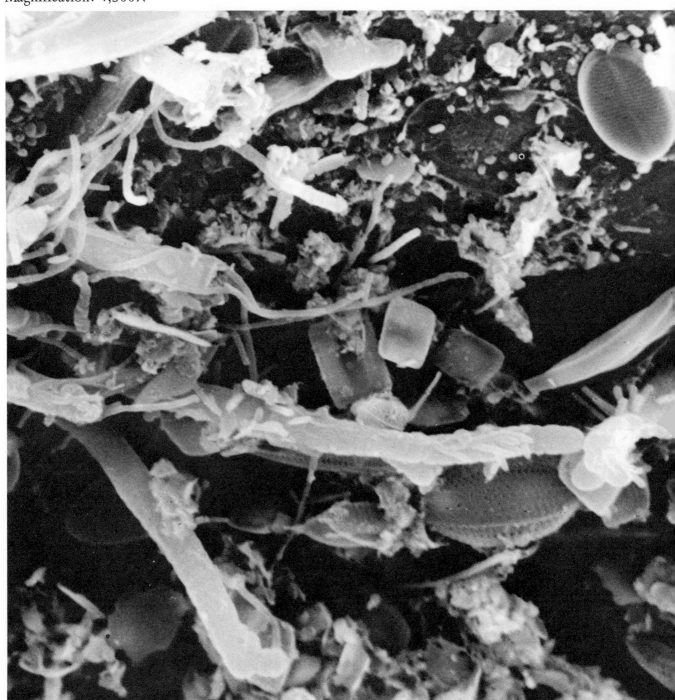

—as well as a variety of bacterial forms.

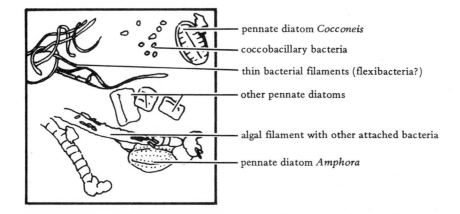

— pennate diatom *Cocconeis*

— coccobacillary bacteria

— thin bacterial filaments (flexibacteria?)

— other pennate diatoms

— algal filament with other attached bacteria

— pennate diatom *Amphora*

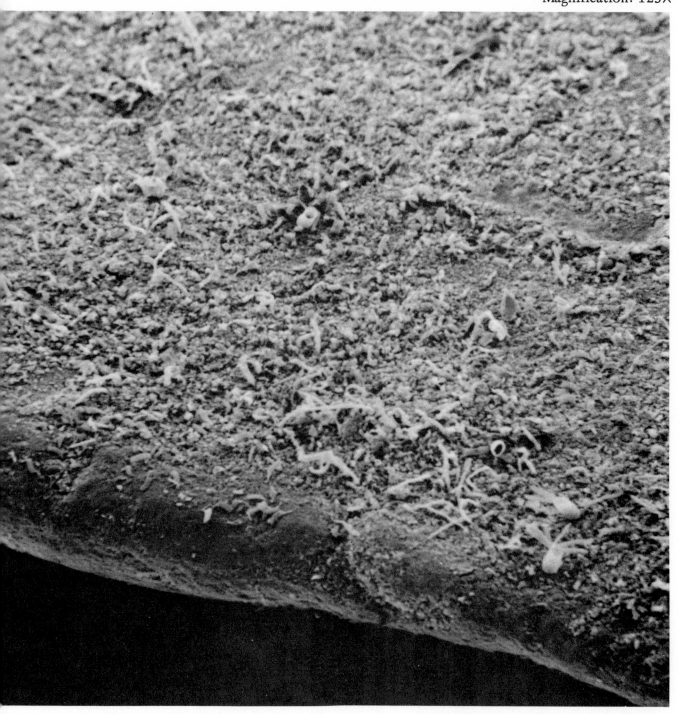

The flat shape of this pebble has apparently prevented it from rolling in the surf, as abrasion is only apparent along its edge—

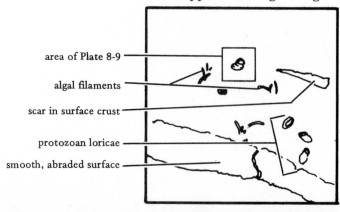

area of Plate 8-9

algal filaments

scar in surface crust

protozoan loricae

smooth, abraded surface

Plate 8-9. Microorganisms on slate pebble
Site: Narragansett Bay, RI
Magnification: 1,275×

—while the exposed top surface is heavily encrusted with debris and microorganisms.

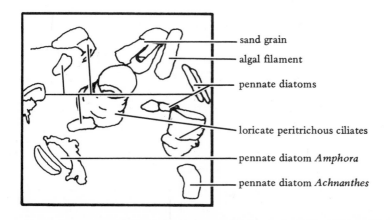

sand grain

algal filament

pennate diatoms

loricate peritrichous ciliates

pennate diatom *Amphora*

pennate diatom *Achnanthes*

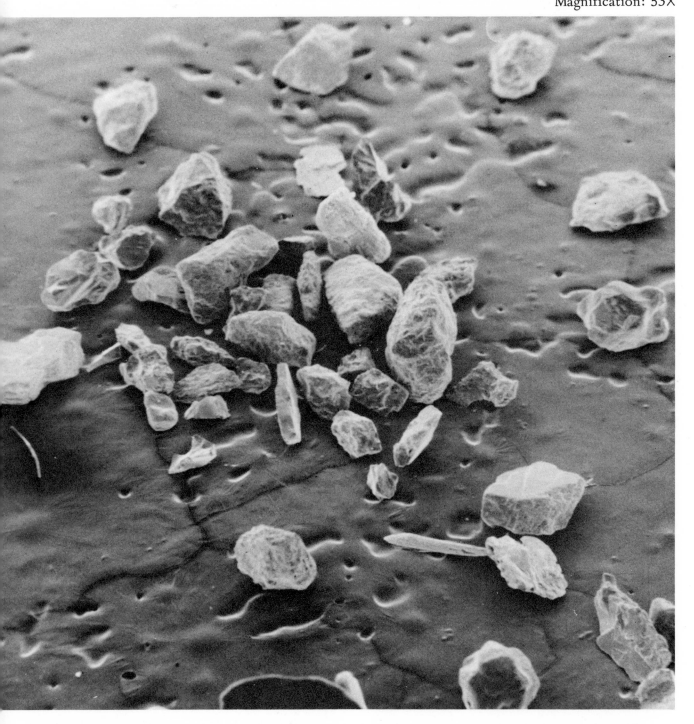

The relatively large sand grains in the surf zone—

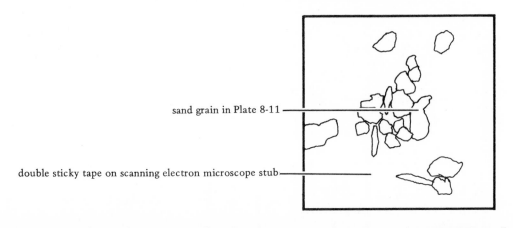

sand grain in Plate 8-11 ⎯⎯⎯

double sticky tape on scanning electron microscope stub ⎯⎯⎯

Plate 8-11. Sand grain from upper surf zone
Site: Narragansett Bay, RI
Magnification: 545X

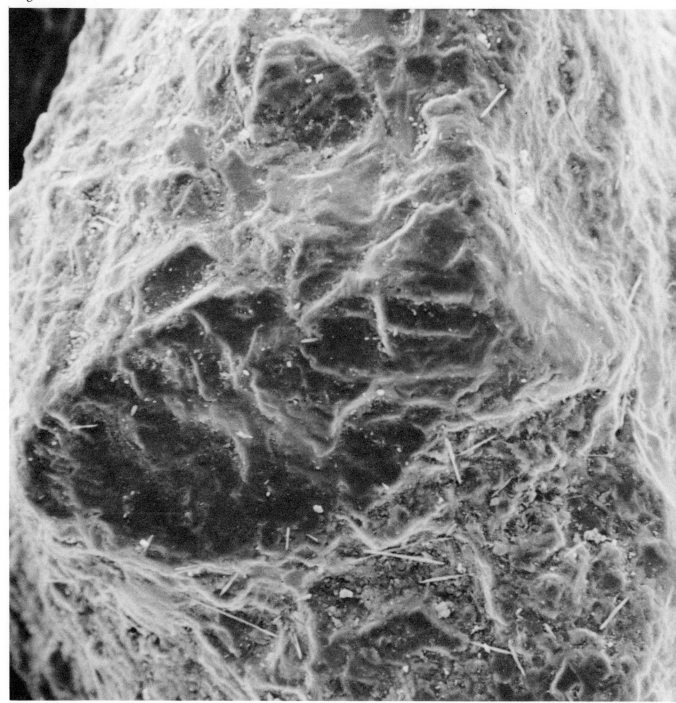

—are quite free of microorganisms—

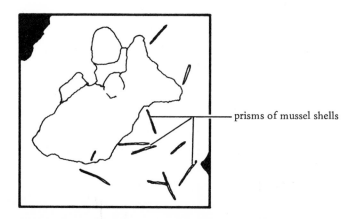

prisms of mussel shells

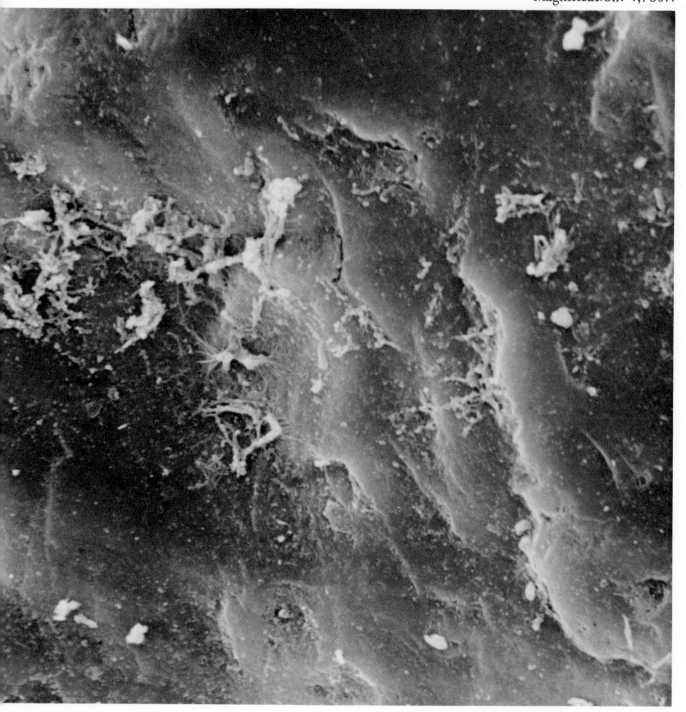

—except for a few tenacious bacteria in some protected areas.

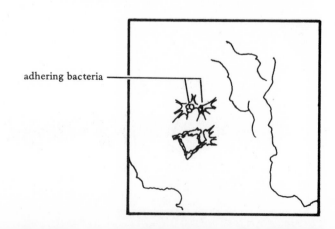

adhering bacteria

*The debris between these sand grains consists of fecal pellets and prisms of mussel shells (*Mytilus edulis*).*

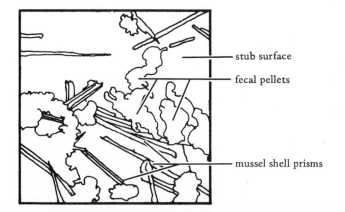

— stub surface

— fecal pellets

— mussel shell prisms

The somewhat smaller sand grains below the surf zone trap some of the finer organic debris.

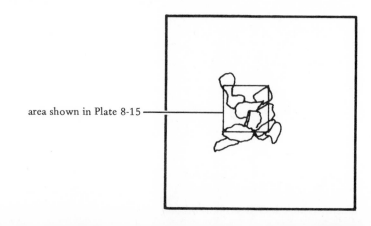

area shown in Plate 8-15

Plate 8-15. Sand grains below surf zone
Site: Narragansett Bay, RI
Magnification: 500X

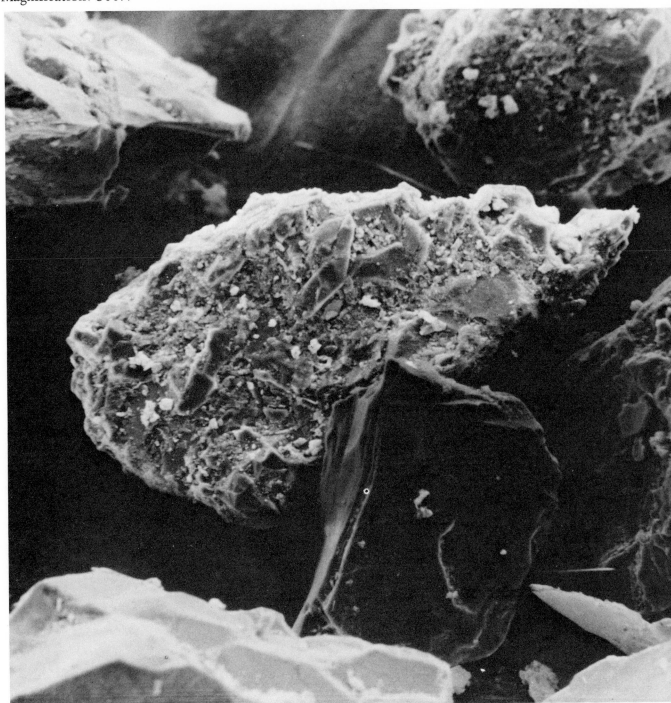

Some have a coarser texture—

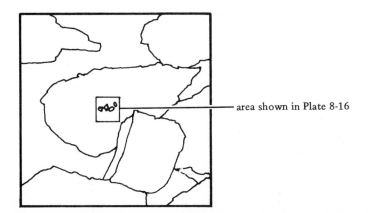

area shown in Plate 8-16

—and are littered with organic debris and diatoms.

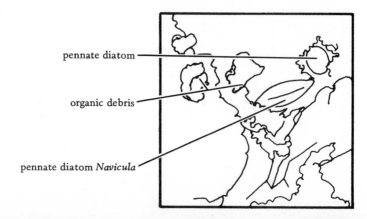

pennate diatom

organic debris

pennate diatom *Navicula*

In quieter and deeper waters the sand gives way to the fine particles of black mud—

Plate 8-18. Particles of black mud
Site: Narragansett Bay, RI
Magnification: 250X

—which consist of small amounts of fine mineral grains and much organic debris enriched with bacterial slime—

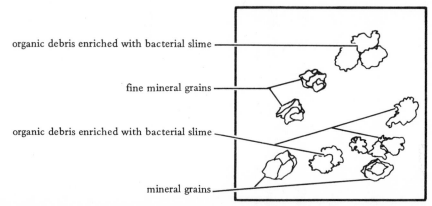

organic debris enriched with bacterial slime

fine mineral grains

organic debris enriched with bacterial slime

mineral grains

Plate 8-19. Particles of black mud
Site: Narragansett Bay, RI
Magnification: 1,246✕

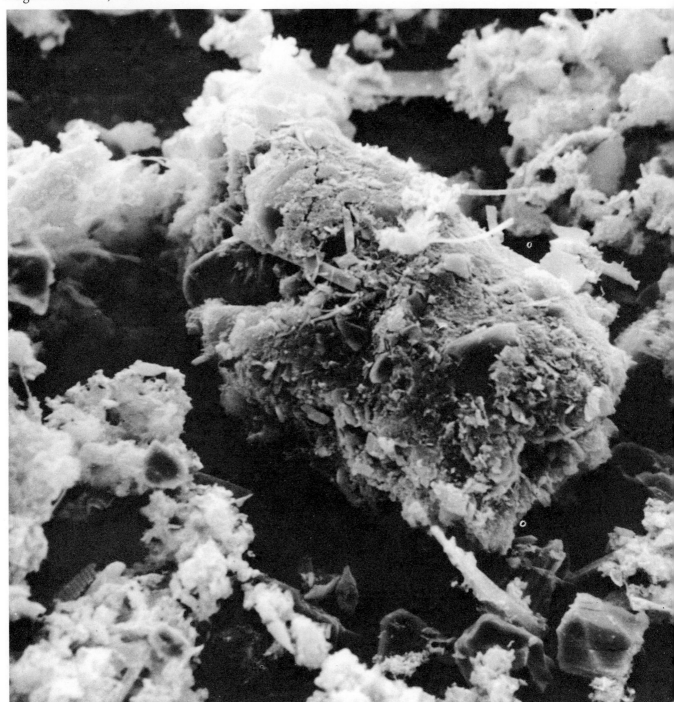

—much of which seems to arise from fragmenting fecal pellets.

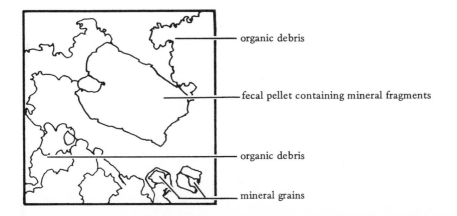

organic debris

fecal pellet containing mineral fragments

organic debris

mineral grains

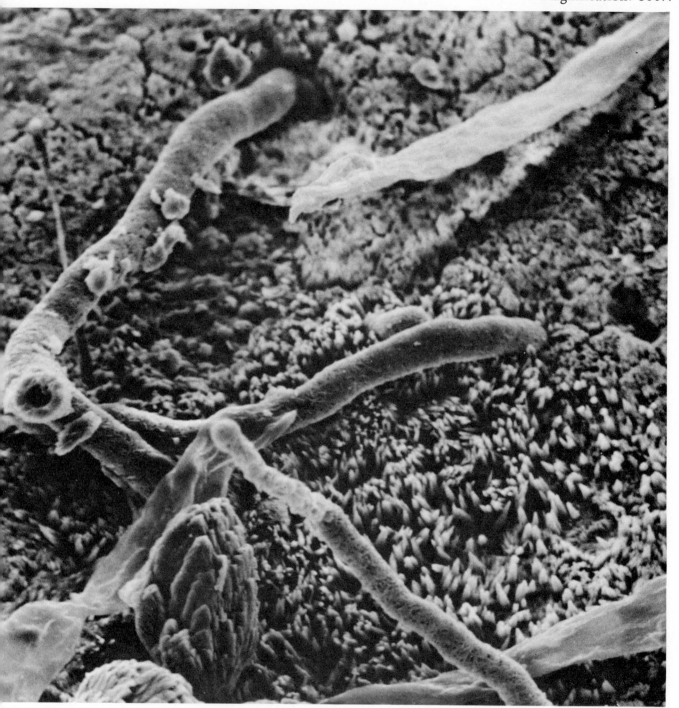

The calcareous skeletons and shells of animals are attacked by boring (endolithic) algae—

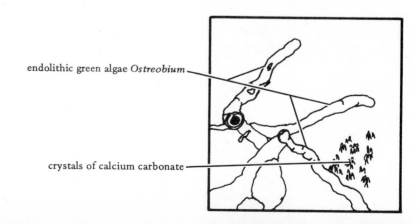

endolithic green algae *Ostreobium*

crystals of calcium carbonate

Plate 8-21. Brain coral skeleton
Site: Sombrero Key, FL
Magnification: 727X

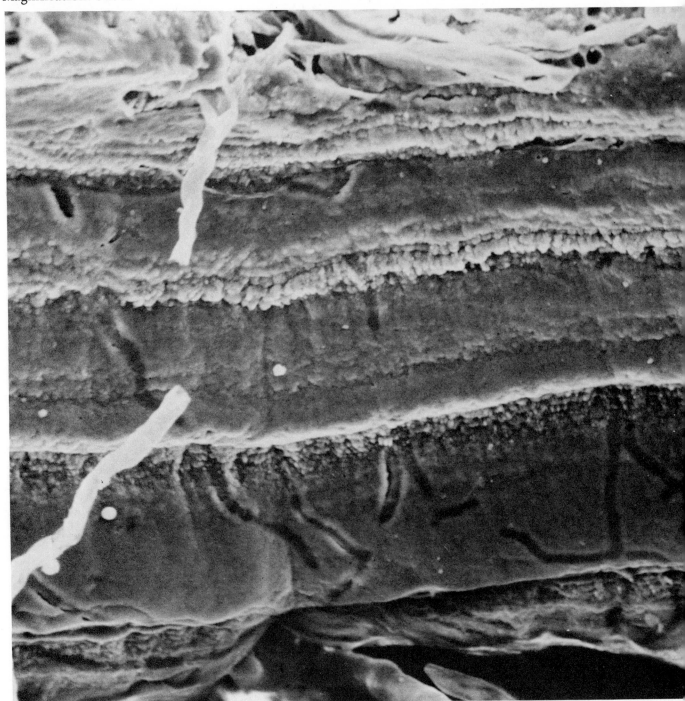

—which leave tunnels through the layers of deposited calcium carbonate.

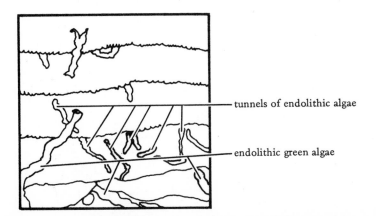

tunnels of endolithic algae

endolithic green algae

Plate 8-22. Deep sea sediment
Site: Atlantic Ocean near the Azores
Magnification: 213X

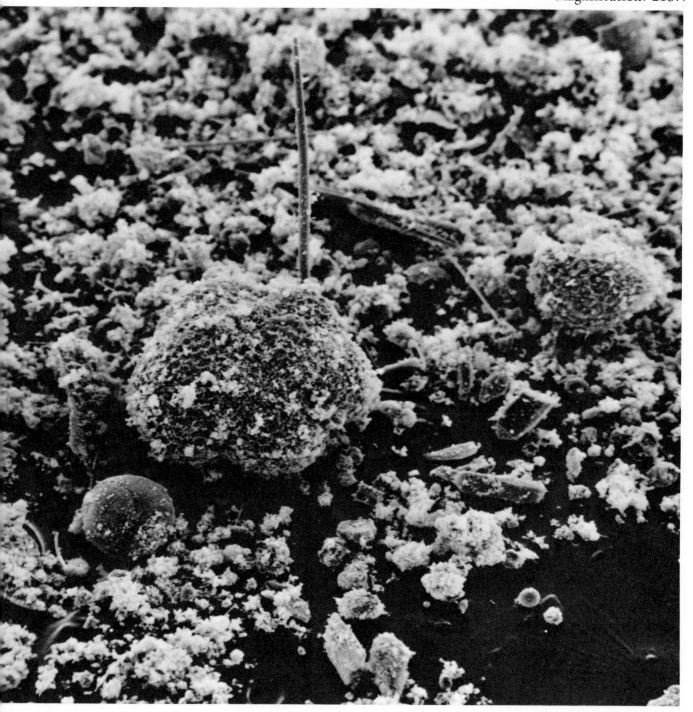

In the sparser sediments of the deep sea, the fine mineral fragments and organic debris are dominated by the shells of planktonic oceanic protozoans such as in this foraminiferal ooze.

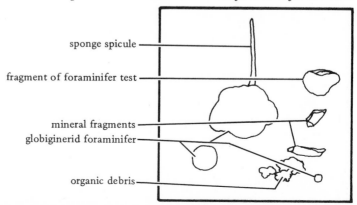

sponge spicule

fragment of foraminifer test

mineral fragments
globiginerid foraminifer

organic debris

Plate 8-23. Juvenile foraminifer test from deep sea sediment
Site: Atlantic Ocean near the Azores
Magnification: 2,900X

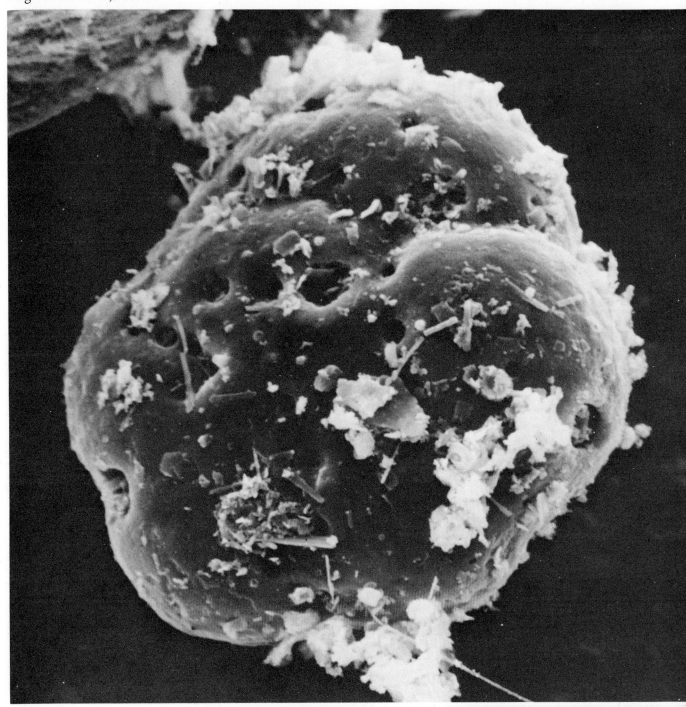

A small test shows the pores through which its inhabitant excretes byproducts; the protoplasmic strands (axopods) used to procure its planktonic food are extruded from the main aperture on the other side.

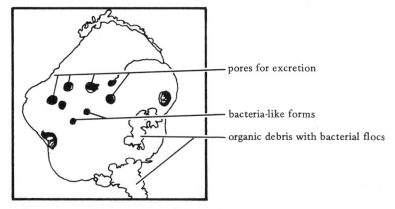

pores for excretion

bacteria-like forms

organic debris with bacterial flocs

Plate 8-24. Foraminifer test from deep sea sediment
Site: Atlantic Ocean near the Azores
Magnification: 1,222X

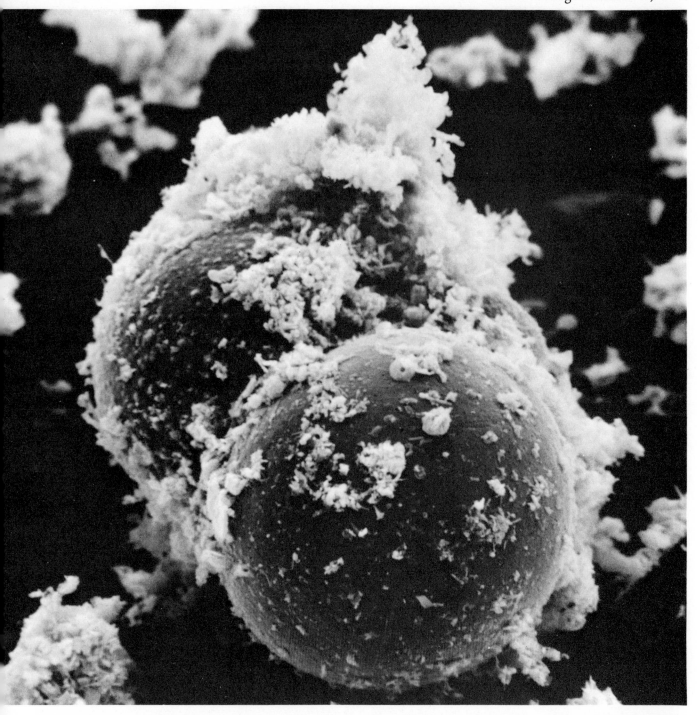

A two-chambered foram test with organic debris and litter from other sedimented materials—

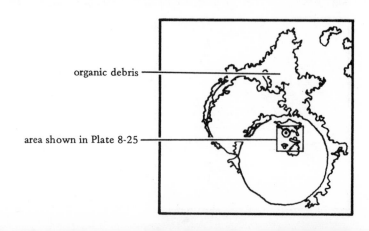

organic debris

area shown in Plate 8-25

Plate 8-25. Sediment debris on foraminifer test
Site: Atlantic Ocean near the Azores
Magnification: 12,525✕

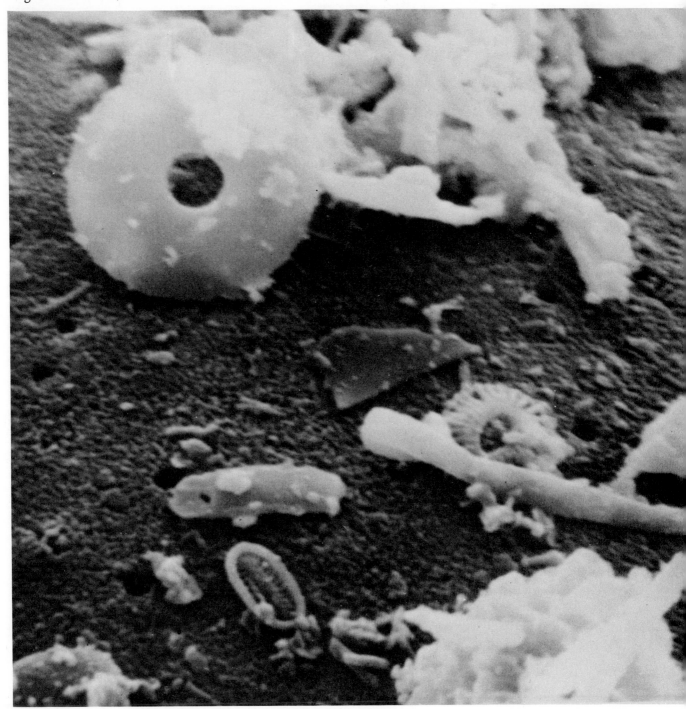

—such as these coccoliths (scales) from coccolithophorid phytoplankton living in the photic zone.

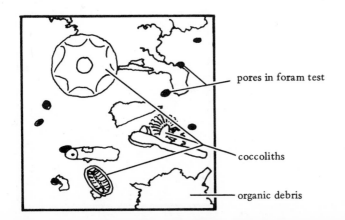

pores in foram test

coccoliths

organic debris

Plate 8-26. Foraminifer tests washed free of sedimentary debris
Site: Atlantic Ocean near the Azores
Magnification: 276X

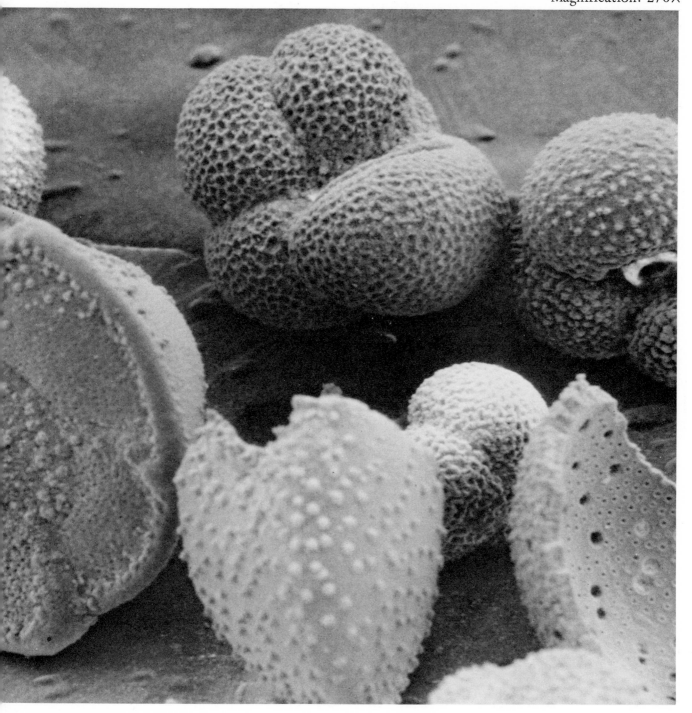

The protozoan remnants such as these calcareous tests of foraminifera—

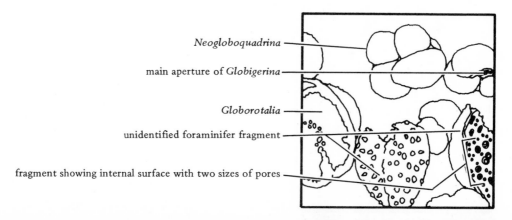

Neogloboquadrina

main aperture of *Globigerina*

Globorotalia

unidentified foraminifer fragment

fragment showing internal surface with two sizes of pores

Plate 8-27. Radiolarian tests washed free of sedimentary debris
Site: Atlantic Ocean near the Azores
Magnification: 525✕

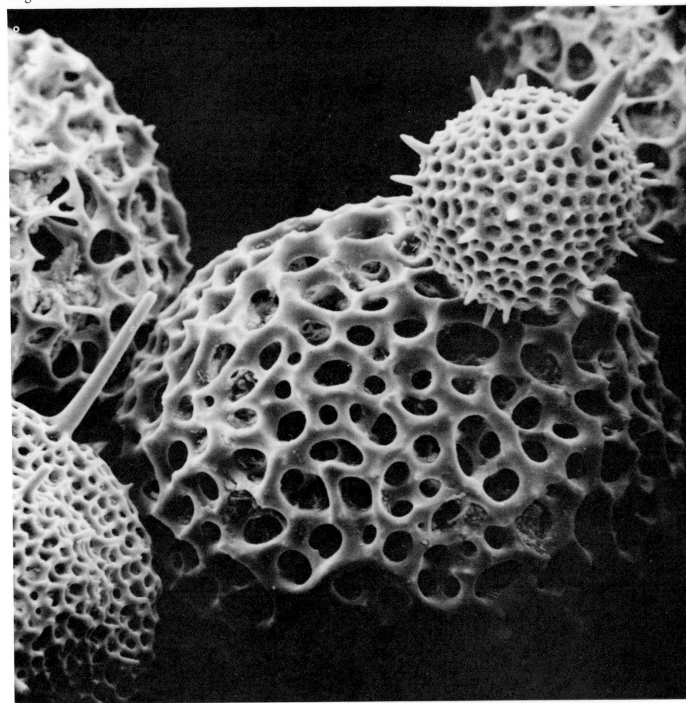

—*and these siliceous skeletons of another group of planktonic oceanic protozoans called radiolarians are used by deep sea geologists to determine the history of the oceans.*

references

Anderson, J. W., and G. C. Stephens. 1969. Uptake of organic material by aquatic invertebrates. Mar. Biol. 4: 243–249.

Arnold, A. F. 1968. The Sea-Beach at Ebb-Tide. Dover Publications, New York. 490 p. (Reprint of 1901 edition by the Century Co.)

Brock, T. D. 1966. The habitat of *Leucothrix mucor*, a widespread marine microorganism. Limnol. Oceanogr. 11:303–307.

Brooks, R. D., R. D. Goos, and J. McN. Sieburth. 1972. Fungal infestation of the surface and interior vessels of freshly collected driftwood. Mar. Biol. 16:274–278.

Bunt, J. S. 1955. The importance of bacteria and other microorganisms in the seawater at MacQuarie Island. Austr. J. Mar. Freshw. Res. 6:60–65.

Carpenter, E. J., and K. L. Smith, Jr. 1972. Plastics on the Sargasso sea surface. Science 175:1240–1241.

Colton, J. B., Jr., F. D. Knapp, and B. R. Burns. 1974. Plastic particles in surface waters of the Northwestern Atlantic. Science 185:491–497.

Conover, J. T., and J. McN. Sieburth. 1966. Effect of tannins excreted from Phaeophyta on planktonic animal survival in tide pools. Proceedings of the Fifth International Seaweed Symposium, Halifax, Nova Scotia, pp. 99–110. Pergamon Press, Oxford.

Crisp, D. J., and J. S. Ryland. 1960. Influence of filming and of surface texture on the settlement of marine organisms. Nature 180:119.

Dodge, J. D. 1973. The Fine Structure of Algal Cells. Academic Press, New York. 261 p.

Droop, M. R., and K. G. R. Elson. 1966. Are pelagic diatoms free from bacteria? Nature 211:1096–1097.

Echlin, P. 1972. Applications to biological materials. *In* J. W. S. Hearle, J. T. Sparrow, and P. M. Cross, (eds.), The Use of the Scanning Electron Microscope, pp. 177–202. Pergamon Press, Oxford.

Fager, E. W., A. O. Flechsig, R. F. Ford, R. I. Clutter, and R. J. Ghelardi. 1966. Equipment for use in ecological studies using SCUBA. Limnol. Oceanogr. 11:503–509.

Fenchel, T. M. 1969. The ecology of marine microbenthos. IV. Structure and function of the benthic ecosystem, its chemical and physical factors and the microfauna communities with special reference to the ciliated protozoa. Ophelia 6:1–182.

Fenchel, T. 1970. Studies on the decomposition of organic detritus derived from the turtle grass *Thalassia testudinum*. Limnol. Oceanogr. 15:14–20.

Fenchel, T. 1972. Aspects of decomposer food chains in marine benthos. Verh. Deutsch. Zool. Ges. 65:14–22.

Frankenberg, D., and K. L. Smith. 1967. Coprophagy in marine animals. Limnol. Oceanogr. 12:443–450.

Gessner, R. V., R. D. Goos, and J. McN. Sieburth. 1972. The fungal microcosm of the internodes of *Spartina alterniflora*. Mar. Biol. 16:269–273.

Haeckel, E. 1887. Report on the scientific results of the voyage of H. M. S. *Challenger*. Zoology, Vol. 18. 1st Part, Porulosa (Spumellaria and Acantharia), pp. 791–888. 2nd Part, Osculosa (Nasselbaria and Phaeodaria), pp. 889–1803. 3rd part, 140 plates.

Hanic, L. A., and J. S. Craigie. 1969. Studies on the algal cuticle. J. Phycol. 5:89–102.

Harold, R., and R. Y. Stanier. 1955. The genera *Leucothrix* and *Thiothrix*. Bacteriol. Rev. 19:49–58.

Hayes, T. L. 1973. Scanning electron microscope techniques in biology. *In* J. K. Koehler, (ed.), Advanced Techniques in Biological Electron Microscopy, pp. 153–214. Springer-Verlag, Heidelberg.

Hendey, N. I. 1959. The structure of the diatom cell wall as revealed by the electron microscope. J. Quekett Microsc. Club 5:147–175.

Hessler, R. R., J. D. Isaacs, and L. Mills. 1972. Giant amphipod from the abyssal Pacific Ocean. Science 175:636–637.

Jannasch, H. W. 1973. Bacterial content of particulate matter in offshore surface waters. Limnol. Oceanogr. 18:340–342.

Jannasch, H. W., K. Eimhjellen, C. O. Wirsen, and A. Farmanfarmaian. 1971. Microbial degradation of organic matter in the deep sea. Science 171:672–675.

Jannasch, H. W., and C. O. Wirsen. 1973. Deep-sea microorganisms: *in situ* response to nutrient enrichment. Science 180:641–643.

Jefferies, R. L. 1972. Aspects of salt-marsh ecology with particular reference to inorganic plant nutrition. *In* R. S. K. Barnes and J. Green (eds.), The Estuarine Environment, pp. 61–85. Applied Science Publ., London.

Johannes, R. E., and M. Satomi. 1966. Composition and nutritive value of fecal pellets of a marine crustacean. Limnol. Oceanogr. 11:191–197.

Johnson, P. W., J. McN. Sieburth, A. Sastry, C. R. Arnold, and M. S. Doty. 1971. *Leucothrix mucor* infestation of benthic crustacea, fish eggs, and tropical algae. Limnol. Oceanogr. 16:962–969.

Jones, E. B. G., and S. K. Eltringham (eds.). 1971. Marine

borers, fungi and fouling organisms of wood. Organization for Economic Co-operation and Development, Paris, 367 p.

Mann, K. H. 1973. Seaweeds: their productivity and strategy for growth. Science 182:975–981.

Margulis, L. 1968. Evolutionary criteria in Thallophytes: a radical alternative. Science 161:1020–1022.

Margulis, L. 1970. Origin of Eukaryotic Cells. Yale University Press, New Haven, Conn. 349 p.

Marsh, G. A. 1970. A seasonal study of *Zostera* epibiota in the New York River, Virginia. Ph.D. Thesis, College of William and Mary, Williamsburg, Va. 167 p.

Meadows, P. S., and J. G. Anderson. 1968. Microorganisms attached to marine sand grains. J. Mar. Biol. Assoc. U.K. 48:161–175.

Meadows, P. S., and G. B. Williams. 1963. Settlement of *Spirorbis borealis* Daudin larvae on surfaces bearing films of microorganisms. Nature 198:610–611.

Melchiorri-Santolini, U., and J. W. Hopton (eds.). 1972. Detritus and its Role in Aquatic Ecosystems. Proceedings of an IBP-UNESCO Symposium. Vol. 29 Suppl. Memorie Dell'Istituto Italiano Di Idrobiologia, Pallanza, Italy. 540 p.

Miner, R. W. 1950. Field Book of Seashore Life. G.P. Putnam's Sons, New York, 888 p.

Moore, H. B. 1931. The specific identification of faecal pellets. J. Mar. Biol. Assoc. U.K. 17:359–365.

Moore, R. C. (ed.). 1954. Treatise on Invertebrate Paleontology. Part D. Protista 3. Protozoa (chiefly radiolaria and tintinnida). University of Kansas Press, Lawrence, Kan. 195 p.

Morita, R. Y., and C. E. ZoBell. 1955. Occurrence of bacteria in pelagic sediments collected during the mid-Pacific expedition. Deep-Sea Res. 3:66–73.

Neushul, M. 1974. Botany. Hamilton Publ. Co., Santa Barbara, Cal. 532 pp.

Parsons, T., and M. Takahashi. 1973. Biological Oceanographic Processes. Pergamon Press, Oxford. 186 p.

Percival, E., and R. H. McDowell. 1967. Chemistry and Enzymology of Marine Algal Polysaccharides. Academic Press, New York. 219 p.

Phleger, F. B. 1960. Ecology and distribution of recent foraminifera. The Johns Hopkins Press, Baltimore. 297 p.

Prakash, A. 1962. Source of paralytic shellfish toxin in the Bay of Fundy. J. Fish. Res. Bd. Canada. 20:983–996.

Pratt, H. L. 1973. MacroSnooping. Skin Diver. Sept. 1973: 38–39.

Pringsheim, E. G. 1957. Observations on *Leucothrix mucor* and *Leucothrix cohaerens* nov. sp. Bacteriol. Rev. 21: 69–76.

Ray, D. L. (ed.). 1959. Marine Boring and Fouling Organisms. University of Washington Press, Seattle. 536 p.

Raymont, J. E. G. 1963. Plankton and Productivity in the Oceans. Pergamon Press, Oxford. 660 p.

Rhode Island Marine Bibliography. 1972. University of

Rhode Island Marine Tech. Rept. no. 3. Kingston, R.I. 192 p.

Robbins, S. F., and C. M. Yentsch. 1973. The Sea is All About Us. 1st Ed. Peabody Museum of Salem, Salem, Mass. 162 p.

Ryland, J. S. 1974. Observations on some epibionts of gulf-weed, *Sargassum natans* (L.) Meyen. J. Exp. Mar. Biol. Ecol. 14:17–25.

Ryther, J. H. 1955. Ecology of autotrophic marine dinoflagellates with reference to red water conditions. *In* F. H. Johnson (ed.), The Luminescence of Biological Systems, pp. 387–414. The American Association for the Advancement of Science, Washington, D.C.

Schrader, H.-J. 1971. Fecal pellets: role in sedimentation of pelagic diatoms. Science 174:55–57.

Sieburth, J. McN. 1966. Recent research sheds light on old paint patent. Maritimes 10(1):15–16.

Sieburth, J. McN. 1968. The influence of algal antibiosis on the ecology of marine microorganisms. *In* M. R. Droop and E. J. F. Wood (eds.), Advances in Microbiology of the Sea, pp. 63–94. Academic Press, New York.

Sieburth, J. McN., and J. T. Conover, 1965. *Sargassum* tannin, an antibiotic that retards fouling. Nature 208: 52–53.

Sieburth, J. McN., and A. S. Dietz. 1974. Biodeterioration in the sea and its inhibition. *In* R. R. Colwell and R. Y. Morita (eds.), Effect of the Ocean Environment on Microbial Activities, pp. 318–326. University Park Press, Baltimore.

Sieburth, J. McN., and C. D. Thomas. 1973. Fouling on eelgrass (*Zostera marina* L.). J. Phycol. 9:46–50.

Sindermann, C. J. 1970. Principal diseases of marine fish and shellfish. Academic Press, New York. 369 p.

Skerman, T. M. 1956. The nature and development of primary films on surfaces submerged in the sea. New Zeal. J. Sci. Tech. Ser. B. 38:44–57.

Taylor, D. L. 1973. Algal symbionts of invertebrates. Ann. Rev. Microbiol. 27:171–187.

Taylor, S. L., and M. Villalard. 1972. Seaweeds of the Connecticut shore: a wader's guide. Bull. no. 18. Conn. Arboretum, New London, Conn. 36 p.

Tootle, J. L. 1974. The fouling microflora of intertidal seaweeds: seasonal, species and cuticular regulation. M.S. Thesis, University of Rhode Island, Kingston, R.I. 50 p.

Tressler, D. K. 1951. Marine Products of Commerce. Reinhold Publ. Corp., New York. 782 p.

Whittaker, R. H. 1969. New concepts of Kingdoms of organisms. Science 163:150–160.

Welsh, B. 1973. The grass shrimp, *Palaemonetes pugio,* as a major component of a salt marsh ecosystem. Ph.D. Thesis, University of Rhode Island, Kingston, R.I. 90 p.

Wiebe, W. J., and L. R. Pomeroy. 1972. Microorganisms and their association with aggregates and detritus in the sea: a microscopic study. *In* U. Melchiorri-Santolini and J. W. Hopton (eds.), Detritus and Its Role in Aquatic

Ecosystems. Proceedings of an IBP-UNESCO Symposium, pp. 325–352. Vol. 29 Suppl. Memorie Dell'Istituto Italiano Di Idrobiologia, Pallanza, Italy.

Wirsen, C. O., and H. W. Jannasch. 1974. Microbial transformations of some ^{14}C-labeled substrates in coastal water and sediment. Microbial Ecol. 1:25–37.

Woods, J. D., and J. N. Lythgoe (eds.). 1971. Underwater Science. Oxford University Press, London. 330 p.

Zischke, J. A. 1973. An ecological guide to the shallow-water marine communities of Pigeon Key, Florida. Seeback Press, Northfield, Minn. 43 p.

microorganism index

surface index

Microbial Seascapes
a pictorial essay
on marine microorganisms
and their environments

design *Sheila Humphreys*

typography *The Composing Room of Michigan, Inc.,
Grand Rapids, Michigan,
in Aldine Roman and Friz Quadrata display type*

printing *Universal Lithographers, Inc.,
Cockeysville, Maryland*

paper *Sixty-pound Cumberland English Finish Offset,
manufactured by the S. D. Warren Company,
Boston, Massachusetts*

**cover printing
and book binding** *The Maple Press Company, York, Pennsylvania*

line drawings *Charles E. West and Sheila Humphreys*

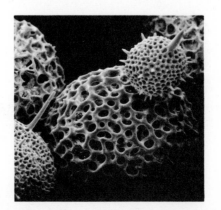